A Natural History of Wine

IAN TATTERSALL AND ROB DESALLE

Illustrated by Patricia J. Wynne

Yale UNIVERSITY PRESS

NEW HAVEN & LONDON

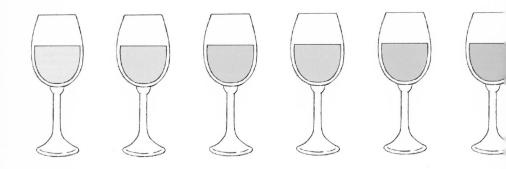

A NATURAL HISTORY OF
Wine

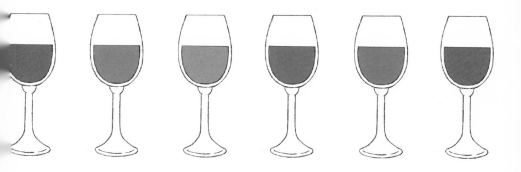

Yale University Press books may be purchased in quantity for educational,
business, or promotional use. For information, please e-mail sales.press@yale.edu
(U.S. office) or sales@yaleup.co.uk (U.K. office).

Designed by Nancy Ovedovitz and set in Caecilia and Torino types
by Tseng Information Systems, Inc. Printed in China.

Library of Congress Control Number: 2015933803
ISBN 978-0-300-21102-3 (cloth : alk. paper)

A catalogue record for this book is available from the British Library.

This paper meets the requirements of ANSI/NISO Z39.48–1992
(Permanence of Paper).

10 9 8 7 6 5 4 3 2 1

To our favorite oenophiles: Jeanne, Erin, and Maceo

Contents

Preface

The book you're holding is the result of an unusual collaboration between a molecular biologist and an anthropologist. As colleagues at the American Museum of Natural History, we have worked together on a number of books on topics relevant to our humanity, including the evolution of the brain and the concept of race. All of this work, of course, involved a lot of hanging out together, and since inspiration is always at a premium when we are writing a book, our head-scratching sessions have tended to be over copious quantities of wine. And once we have started drinking wine, at least a good one, the conversation naturally tends toward what we happen to be imbibing. Wine is like that: it appeals so comprehensively to the senses that it has trouble staying in the background. All those dreadful wine-and-cheese receptions notwithstanding, in its better incarnations wine is anything but a wallpaper beverage, and learning about it should not be the self-conscious, joyless burden portrayed in the movie *Somm*. It should at once be a fascinating, satisfying, and above all relaxing experience.

As we talked, we realized that wine holds a place in virtually every major area of science—from physics and chemistry to molecular genetics and systematic biology, and on through evolution, paleontology, neurobiology, and ecology, to archaeology, primatology, and anthropology. We also came to realize how much knowing about this complex beverage—what it is, where it comes from, and how we respond to it—enhances our enjoyment of it. Hence this book, the initially unintended product of the many conversations we have had about wine, which, it turns out, has many more dimensions than either of us had suspected previously.

Of course, a book like this one is never written in a vacuum. We have benefited over the years from interactions with many wine professionals and oenophiles. On the professional side, among many others we would like especially to acknowledge our debt to Patrick McGovern, the preeminent authority on ancient wine and its composition, and to Rory Callahan, whose knowledge of wines worldwide is as vast as its dispensation is understated. Among amateur oenophiles (in the literal sense of that term), we would like to express our particular appreciation to Neil Tyson, Mike Dirzulaitis, and Marty Gomberg, all of whom have introduced us to many fabulous wines to which we would otherwise never have had access. Vivian Schwartz and Jeanne Kelly were kind enough to read and comment on the manuscript, as, very usefully, were three anonymous reviewers. The book itself could never have come about without the enthusiastic support and input of Jean Thomson Black, our editor at Yale University Press, and the patience and forbearance of Samantha Ostrowski, who shepherded it through the production process. We are indebted also to our manuscript editor, Susan Laity, who rigorously tightened up our text; and on the visual side our immense gratitude goes to Patricia Wynne, illustrator par excellence, with whom it has, as always, been a pleasure to work. We also thank Nancy Ovedovitz for the book's elegant design.

A Natural History of Wine

1

Vinous Roots

Wine and People

COLLECTION WINE

ARENI COUNTRY
RED DRY TABLE WINE
2005
ԱՐԵՆԻ ԳՅՈՒՂԻ
ԿԱՐՄԻՐ ԱՆԱՊԱԿՉԻՆ

The label on the bottle didn't look like much. "*Areni Country Red Dry Table Wine,*" it said. On a wintry morning in New York City our expectations were, frankly, not too high for this obscure wine, produced in a tiny village in a remote corner of Armenia. So imagine our delight when it leapt from our glasses, all bright red fruit and black cherries, with just enough texture to leave a lingering memory that made us eager for more. Even better, it had been produced a mere kilometer or so up the road from the place where wine arguably began.

To find the world's oldest winery, leave the Armenian capital of Yerevan behind you in the shadow of looming Mount Ararat and drive fast for two hours southwest. The meandering road will take you across some pretty unforgiving terrain, part of the harsh and rugged volcanic plateau that lies at the foot of the Lesser Caucasus Mountains. This is hardly the most promising of territories for an oenophile, and before too long you may find yourself despairing of ever spotting a vine amid the waste of short brown grass and eroded hillsides that stretches to the horizon in all directions. But after a while, a small green oasis will open up in front of you: a cluster of orchards and vineyards and beehives, all given life by a narrow, chattering river that seems to spring from nowhere. In the center of this lonely agricultural outpost lies the village of Areni, a small cluster of buildings that is mostly hidden by the lush vegetation surrounding it. And although the village itself is as obscure as it is tiny, its name is not. You'll see that name on bottles of wine sold all over Armenia, not because Areni itself is nowadays a major wine producer but because centuries ago this little village gave its name to what some consider Armenia's finest wine grape.

Almost every wine-drinking visitor to Armenia will at some point taste a bottle or two of the deep-ruby Areni wine; and if that visitor is fortunate enough to get a really good one, its light fragrance, backed up by a firm texture, lingering ripe plum and dark cherry flavors, and, in the best of them, a hint of black pepper in the finish, will not soon be forgotten. Even an ordinary Areni is typically delicious on a hot day, poured straight from a jug kept in the refrigerator and preferably enjoyed while relaxing in the shade of one of the overflowing grape arbors with which Armenia is so generously endowed. But the people of Areni village know their wine, and they know their grape, and they probably correctly doubt that it can be grown to the same advantage anywhere else. After all, they will tell you, they have been growing these vines for centuries—for so long, indeed, that the memory of winemaking in these parts fades back into the mists of time.

The earliest cuneiform inscriptions referring to wine production in Greater Armenia date to the days of Urartu, a proto-Armenian kingdom centered in eastern Anatolia that flourished in the seventh and eighth centuries B.C.E. Urartu was a major exporter of wine to neighboring Assyria, and most Urartian cities had important wine-storage facilities, some holding many thousands of liters, which testify to the beverage's economic importance. The first literary records of wine in the region come from the beginning of the fourth century B.C.E., when the soldier Xenophon, in his epic work *Anabasis,* described the retreat of a Greek mercenary army from Babylonia. Xenophon records that, as they fought their way across southern Armenia on their way to the Black Sea, the Hellenic forces "took up their quarters . . . in numerous beautiful buildings, with an ample store of provisions, for there was wine so plentiful that they had it in cemented cisterns." As ancient as Xenophon's account and the wine jars of Urartu may be, though, at Areni the story of wine began immensely earlier yet. For in a cave just outside the village, archaeologists have found traces of winemaking that probably date from a full six thousand years ago.

Past the bucolic Areni settlement, the scenery changes dramatically. As you leave the fertile valley behind you'll enter a narrow chasm carved by the river Arpa through a massive outcropping of limestone. And low on the sheer cliff to your right, just before the river is joined by a tributary running down an equally precipitous gorge, there opens the entrance to the cave that is now famous to archaeologists as Areni-1. First mapped in the 1960s by Soviet Cold War planners improbably on the lookout for places to shelter the sparse local population from nuclear attack, Areni-1 has since proven a bonanza for prehistorians, its extraordinary archaeological richness stemming from the many advantages it has offered people throughout history. Not only is the cave roomy and strategically located above the valley, but its arching portal made it a comfortable place for early humans seeking shelter from the elements. What's more, in later times the cave's interior provided ideal conditions for the preservation both of the dead and of the artifacts they used in life.

You can park your car in the shade of a sprawling grape arbor beside the bickering Arpa and scramble up a steep talus slope toward the cave entrance: a high, wide gash across the side of the cliff. As the narrow path flattens out onto the platform of sediment at the cave's mouth, you'll glimpse a partially excavated area in which archaeologists have already found hints of almost unimaginably long-term use of the cave by humans. At the bottom of the pile of occupation deposits, test pits have produced crude stone tools indicating that Ice Age hunter-gatherers were camping at Areni some hundreds of thousands of years ago, long before our species *Homo sapiens* came into existence. Doubtless these early humans were exploiting the rich bounty of a local environment that would have included fish teeming in the river as well as the herds of migrating mammals that converged down the neighboring valleys. It is not hard to imagine those early human relatives perched outside the cave, scanning the river valley for approaching prey.

For now, though, the imagination must serve. Sadly for those interested in Ice Age lifeways, it will be a good while before we are able to say much more about the earliest inhabitants of Areni-1—though for the best of reasons. Because this place was always attractive to humans, the oldest layers at Areni are covered over by the remains of more recent occupations that will have to be painstakingly documented and removed before the Ice Age layers are reached.

Still, the archaeologists are in no hurry, for in the more recent layers they have been finding an unparalleled record of life in the crucial interval, some six thousand years ago, between the New Stone Age and the Bronze Age. This was a time when complex settled lifeways were just becoming established in the Near East, and when Areni-1's Chalcolithic (Copper Age) inhabitants were, among other things, burying their dead in the cave's gloomy interior.

As you leave the light and airy platform at the front of the cave and proceed deeper into the rock, the natural illumination is gradually replaced by a string of feeble lightbulbs that reveal a tall, winding passageway bounded by a deep pit on the left. A sharp right turn, through what is in effect a natural airlock, brings you to a wider section of the cave, in which shallow excavations into the floor have revealed evidence of several extensive Chalcolithic occupations.

What makes the findings at Areni-1 special is that the cool, dry conditions beyond the airlock proved ideal for the preservation of light organic materials of the kind that usually rot away rapidly and disappear. Rarities of preservation include pieces of rope, textiles, and wooden implements—even a complete shoe, made from a single piece of leather. This remarkable artifact caused quite a stir when it was first reported, not only because of its excellent preservation but also because of its age: in the entire archaeological record of the Old World, only the damaged shoes worn by Ötzi the Ice Man, the natural mummy of a Chalcolithic hunter discovered in 1991 after an Alpine glacier melted, come anywhere close to its antiquity. And Ötzi's grass-stuffed shoes are several hundred years younger than the Areni moccasin.

Equally remarkable at Areni-1, though, is the extent of the evidence about their everyday existence that the ancient people left behind. Within the shelter they built dwellings with durable walls and smooth, plastered floors; they cooked food over hearths; they made tools of obsidian and chert; they ground grains on flat stones—and they made wine. Indeed, the Chalcolithic people of Areni-1 have bequeathed to us the remains of the world's earliest winery: the first tangible physical evidence we have, from anywhere, of a society's devotion to the fermented juice of the grape.

In 2007, archaeologists were carefully removing the superficial occupation debris that had accumulated in the cave when they found their way down to a layer that revealed a shallow, flat-bottomed basin with raised edges, scraped into the hard-packed clay of the ancient cave floor. The bot-

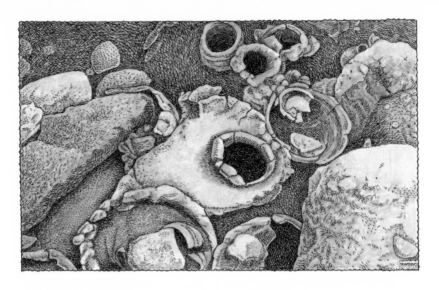

The grape-treading platform at Areni, with the sunken receptacle at its lower end (*center*) and other wine jars surrounding it (after a photo by Boris Gasparyan)

tom of this basin sloped slightly, toward the mouth of a large (60 liter) pottery jar that was sunk into the cave floor beside it. The scientists at once recognized the flat platform as a surface on which ancient grapes had been trodden (presumably by unshod feet). The juice had drained naturally into the jar, which had clearly served as a fermentation vat. The cool, dry conditions of the cave would have provided a perfect environment for the fermentation process, as well as for the wine's later storage in the many other pottery jars lying in the immediate vicinity. The purpose of this unusual archaeological feature was apparent from the start, not only because of its resemblance to wineries known from later times but because the treading area was littered with grape seeds and stems from a strain of today's favored winemaking vine species, Vitis vinifera.

This staggeringly old winery was an exciting find, especially given the sophisticated arrangement of the pressing floor and the large size of the fermentation vat. Usually scientists who try to find traces of the very early production and consumption of wine have to make do with more indirect evidence, most notably the chemical residues that form on the insides of containers used to store the wine. The study of such residues has an intriguing history in archaeology (colleagues recently found traces of marijuana inside Elizabethan clay pipe stems dug up in William Shakespeare's garden in Stratford-upon-Avon), but the evidence is sometimes not easy to interpret. A handful of potsherds found on the grape-pressing floor at Areni-1, for example, radiocarbon-dated to between 6,100 and 6,000 years ago, proved to carry residues of malvidin, a major pigment in grape skins that is responsible for the color of red wine. A wonderful discovery, but a bit equivocal nonetheless: malvidin is also present in fruits other than grapes, such as pomegranates, which still grow around Areni today.

Because the source of the malvidin might not have been grapes, Patrick McGovern, the leading expert on analyses of this kind, observed that he would have been more comfortable if traces of tartaric acid had also been found on the pottery fragments. For unlike malvidin, in the Near Eastern environment tartaric acid is a compound pretty much confined to grapes. Still, given all the supporting evidence that the structure at Areni-1 was a winery, it seems reasonable to conclude that the malvidin came from wine. And though the discovery itself was serendipitous, finding a winery this old was not hugely surprising since several years before the Areni discovery was made, McGovern himself had reported tartaric acid residues from the remains of a pottery jar found at an even earlier site, Hajji Firuz Tepe in the Zagros Mountains of Iran.

The jar in question was manufactured at some time between about 7,400 and 7,000 years ago, and additionally bore traces of resin from the terebinth tree. It is plausible to conjecture that this resin was added to preserve the wine in the container, and it would probably have made the resulting drink taste rather like Greek retsina. The practice of preserving wine with resin is documented well back into classical times, and most authorities reckon the tradition started a lot earlier than that. And while it remains possible that the resins might have been used simply to seal the unglazed pottery, the traces of resin at Hajji Firuz Tepe suggest that the wine stored in the jar was deliberately made, rather than produced by the accidental fermentation of grape juice.

✦ ✦ ✦

Perhaps a little historical perspective will be useful at this point. At about seven thousand years old, the mud-brick settlement at Hajji Firuz

Tepe dates from late in the Neolithic (New Stone Age) period. The Neolithic was the time, following the final retreat of the northern glaciers at the end of the last Ice Age, when the Near East was the site of the first human experiments in settled life based on the domestication of plants and animals. By the beginning of the Neolithic human beings who looked anatomically just like us had already been on earth for well over 150,000 years, and the modern human creative spirit had been burgeoning for at least half that time. In Africa, early stirrings of the modern mind have been detected back to around one hundred thousand years ago, and the earliest of the fabulous cave paintings of France are well over thirty thousand years old. But even the geniuses who decorated caves such as Lascaux and Chauvet were still practitioners of an ancient hunting and gathering lifeway that had its roots in a still more distant past and ensured that human beings remained relatively thin on the landscape. As a result, in economic and social terms the Neolithic represented by far the most fateful innovation in all of human prehistory. Settled life in villages—and soon in towns and cities—represented a complete break with the past: the greatest revolution ever in the relationship between human beings and the world around them.

Until the end of the Ice Ages, human beings had lived off nature's bounty and by its rhythms. But as the northern ice caps started to retreat in earnest around eleven thousand years ago, people in several centers around the world began to experiment with permanent settlement based on agriculture. The Syrian site of Abu Hureya is particularly instructive: it records a transition that ran from hunting and gathering between 11,500 and 11,000 years ago, through hunting and gathering supplemented by cereal cultivation about 10,400 years ago, to both plant and animal domestication—still augmented by hunting and gathering—by around 9,000 years ago. A fully settled way of life was the apparently inevitable outcome of such developments, and once this had been achieved the tempos of both social and technological change began to accelerate. Towns protected by walls began to appear in the Near East by about 8,500 years ago, and it then took a mere 3,000 years for complex stratified urban societies to become well established in the region.

Hajji Firuz Tepe itself was a village of modest size, but it existed at a time of rapid economic and doubtless also social change; only a thousand years later, the winery at Areni-1 was more or less contemporaneous with the first stirrings of the urban Sumerian civilization in Mesopotamia to its south. Both sites, though, document times when the manufacture of pottery had long been a feature of life in the Near East, whereas earlier sites such as Abu Hureya belonged to the Pre-Pottery Neolithic, a period during which settled lifeways were being adopted, but ceramic technology had not yet been invented. All in all, we can surmise pretty confidently that by Hajji Firuz Tepe times a tradition of winemaking was already well established in the Near East, although whether wine is the earliest fermented beverage ever made is less certain.

One reason for this uncertainty is that among the first plants to be domesticated in the Old World were the cereals: wheat and barley in western Asia and rice in China. The deliberate production of fermented beverages clearly followed closely on cereal domestication; in China there is trace evidence in the eastern province of Henan for the production of "beer" (fermented from rice, honey, and fruit including grapes) by about nine thousand years ago. Probably the early Neolithic of the Near East was a time of similar experimentation based on local cereals—certainly after pottery had become available. Indeed, occasional argument still erupts among scientists over whether the first cereal product in the region was beer or bread.

Yet it may be significant that gathering grapes (or any other fruit) and fermenting their juice is a less complex process than doing the same thing with cereals, in which laborious intervention is needed to convert the starches into sugars before fermentation can proceed. Quite simply, it's easier to make wine than beer—after all, nature does it unaided. And grape seeds found at Abu Hureya show that very early on, even before pottery containers were available to potential winemakers, people in the Near East were taking an interest in the fruit of the vine. At the same time, though, it may be significant that whereas beer was being made in China very early indeed, the first evidence for Chinese viticulture goes back only about 2,300 years. Significantly, evidence for this comes from far-flung Xinjiang, where influences from western Asia would have penetrated first along the precursor to the Silk Road trading network. Possibly the rice-centered culture of Han China, with its range of cereal-based alcohols, was already set in its habits and resisted competition from the grape.

Whatever may have been going on in eastern Asia, investigations into the origins of grape growing and winemaking have repeatedly converged on the western fringes of the continent. Recent DNA studies produced results compatible with the notion that vines were first domesticated in the southern Caucasus, and although this origin is not conclusively proven, wine and its associated rituals seem even more deeply ingrained in western Caucasian society than in other famously oenophilic countries such as France. Certainly, no visitor to Georgia and the vine-friendly parts of Armenia can fail to be impressed by the importance of wine in hospitality rituals, and by how thoroughly vine-festooned the houses are in every country village.

There is plenty of archaeological evidence that viticulture had spread widely to Mesopotamia, the Jordan valley, and Egypt by about five thousand years ago. Even if the geographical dispersal of grape growing out of the Caucasus was relatively rapid, it would have taken some time for it to cover such a wide swath of territory, making it easy to imagine that the initial domestication occurred in the Caucasus early in the Neolithic, perhaps a couple of thousand years before the winery was constructed at Areni-1, or conceivably earlier yet. Whether vines were initially domesticated—at a time when plant domestication was all the rage—in order to provide a supply of fleshy table grapes rather than juice for fermentation is something that can be endlessly debated. But what is clear is that one use of the vine would inevitably have followed closely upon the other.

One intriguing aspect of the winery at Areni-1 is its location within a "cemetery" of wine jars that had also served as urns. These burial jars contained the remains of several individuals of various ages, and although the remains of the men had been cremated, those belonging to women and juveniles had been dismembered. Drinking cups made of animal horn were also found in and around the interments. Boris Gasparyan, the lead excavator at Areni-1, believes that there was a close relationship between the winemaking and the activities associated with the cremations, dismemberments, and burials. If so, the extraordinary site of Areni-1 inaugurated a tradition of using fermented beverages in funerary and other rites that is abundantly documented in later phases of antiquity.

Such codification of alcohol use speaks directly to the inherent human tendency to give symbolic meaning to experience and ritualize behaviors of all kinds—perhaps especially those involving altered physiological states. From the earliest times, wine certainly had the supremely practical utility of easing social tensions in addition to the more symbolic but equally functional purposes of cementing reciprocal relationships and lubricating social rituals. And almost equally certainly, wine sometimes also had its place in shamanistic and other rites. It's easy to imagine that early hominids also occasionally got drunk on naturally fermenting fruit, and it is even possible that early modern hunters and gatherers devised ways of fermenting honey or fruit juices before—perhaps long before— pottery containers were invented. But adapting such practices to a ritualistic context is unique to modern humans, as we still see today in the consumption of sacramental wine or even in such unfortunate ritualistic expressions as the Saturday-night binge drinking of soccer hooligans.

✦ ✦ ✦

In the Republic of Georgia, where some country winemakers still ferment their grape juice in *qvevri*, large buried clay pots that are direct descendants of their smaller counterparts at Areni, wine is as deeply embedded in the human spirit as might be expected of one of the regions where the beverage may have originated. Tradition there still dictates that the host or guests at a feast choose a toastmaster (the *tamada*) to preside over the wine drinking. Well-developed social skills and cleverness with words are the keys to this role. The feast unfolds as the proposal and answering of a succession of elaborate and often witty toasts, honoring everything from the nation's glory to present and absent friends and relatives. After each toast, all present drain their glasses, and although theoretically no one drinks between toasts, the revelers can become pleasantly tipsy over the numerous courses of the meal. But just spare a thought for the poor tamada, who will have emptied many a glass before the feast is finished, but is expected to show no symptoms of inebriation. Modern Georgian rituals happily demonstrate that tradition and enjoyment need not conflict in the consumption of wine. But they remind us also that in the ancient world, as in ours, wine drinking was often bound up with rules, rituals, and cosmic beliefs, perhaps most especially in places where it was an expensive import.

Wine was prized from ancient times as a beverage of status and ostentation. At about 3150 B.C.E. Scorpion I, a predynastic king of Upper Egypt, was laid to rest in a many-chambered tomb, three entire rooms of which were filled from floor to ceiling with jars that now contain grape seeds, the chemical residues of wine, and the terebinth resin sealant with which we are already familiar from Hajji Firuz Tepe. Figs had apparently been added to some jars to improve the wine's flavor, or perhaps to provide yeast or sugars to aid in fermentation. All together there were about seven hundred jars in the three rooms, together containing close to four thousand liters of wine—more than enough to give Scorpion a splendid start in the afterlife. The wine itself turned out to have been shipped from the southern Levant, on the western coast of the Mediterranean many hundreds of kilometers away, although the jars may have been locally resealed as part of Scorpion's funeral rites.

The evidently oenophilic Scorpion was far from the only Egyptian with a pronounced taste for wine. At Saqqara, site of the great Step Pyramid of Djoser, an inscription dated to about 2550 B.C.E. records that Metjen, an official of the pharaonic court, made a "great quantity of wine" in a walled vineyard, probably located in the Nile Delta, where temperatures were moderated by the proximity of the Mediterranean. And as remote in time as they are from us, the Egyptians developed many conventions that we think of as modern. Once wine production in the delta had become established, they rapidly devised what amounted to a classification system, analogous to the rankings and appellations developed in France thousands of years later. Individual wine containers were labeled with the name of the region, the year of production, and even the name of the winemaker. The most fortunate producers were identified as makers of wine for the pharaoh. Wines might be unclassified or ranked as "genuine," "good," or "very good."

It became rapidly de rigueur for wealthy Egyptians not only to be washed with wine before being mummified but also to be buried with a

Three scenes from the walls of New Kingdom tombs at western Thebes, Egypt: (*top*) harvest scene from the Tomb of Khaemwaset (19th Dynasty); (*below*) scenes from the Tomb of Userhat (19th Dynasty); (*left*) wine jar inscribed, "Wine of Lower Egypt for the deceased Lady Nodjmet" (18th Dynasty)

selection of the finest wines. Soon this custom became rigorously codified, and by around 2200 B.C.E. it had become unthinkable among Egypt's elite not to be buried with wines from the five most prestigious regions of the Nile Delta. Just as today, when wine has become a fashion accessory and an investment vehicle, in ancient times it seems that some of the best wine never got drunk! Still, the Egyptians were as pragmatic in this regard as in others; if the wines themselves were not available, or were unaffordable, it eventually became enough to illustrate, or even merely to list, them on the tomb walls.

One specific use for wine in ancient Egypt was in curing diseases. The alcohol in wine makes it an excellent vehicle for dissolving ingredients such as resins and the compounds present in medicinal herbs. Wine is hence an ideal medium for delivering medications to the sick, and written records show that as long ago as 1850 B.C.E. herbal wine infusions were prescribed in Egypt for afflictions as diverse as stomach problems, respiratory conditions, constipation, and herpes. What is more, molecular archaeology reveals that the qualities of wine as a healing medium were almost certainly recognized by the Egyptians much earlier than this. Chemical analysis has demonstrated that one of the jars placed in Scorpion's tomb almost 5,200 years ago contained a cocktail of herbs that included balm, coriander, senna, mint, and sage. It is a good bet that the purpose of this complex mixture was medicinal—which makes us wonder about the kind of afterlife Scorpion expected to have!

Still, elaborate as his funerary ceremonies doubtless were, Scorpion does not hold the record for early ostentation. In 870 B.C.E., Assurnasirpal II of Assyria held what was probably the most epic bash ever at his new capital of Nimrud, in the northern Tigris valley. In ten days of feasting, about seventy thousand guests consumed ten thousand skins of wine, in addition to two thousand cows and calves, twenty-five thousand sheep and lambs, several thousand birds, gazelles, fish, and eggs, and more. Other beverages included ten thousand jars of beer, each containing several liters, roughly as much as a wineskin. Significantly, when the king is depicted feasting in the commemorative bas-reliefs at Nimrud, he is not shown drinking the beer that was almost emblematic of Assyrian society (and was, indeed, the medium in which Mesopotamian workers had typically been paid since at least 3400 B.C.E.). Instead, Assurnasirpal is shown brandishing a wine bowl.

The Greeks seem to have benefited early on both from Egyptian winemaking proficiency and from the transportation advances of the Levantine Canaanites, whose cedar of Lebanon ships pioneered the long-distance transport of wine around the Mediterranean. They built on this maritime expertise to become the first to produce wine on a truly commercial scale and to turn the beverage into a commodity available to virtually all. From one single Greek merchantman, wrecked off the Mediterranean coast of France in the fifth century B.C.E., underwater archaeologists in the twentieth century recovered a full ten thousand amphorae that had contained the equivalent of more than three hundred thousand modern bottles of wine. From a variety of literary sources, we know that the Greeks learned how to concentrate the sweetness of grapes by drying them on mats before crushing them, and to harvest them early to preserve their acidity. In addition, they developed their own wine-drinking decorum: in contrast to the *barbaroi*, who drank their wine straight, the Greeks watered their wine, often in the formal setting of the *symposion*. But water was not the only adulterant used in wine: laws governing the labeling of wines are evidence of rampant fakery, as a preference for older wines developed and different regions strove to distinguish themselves by packaging their product in amphorae of distinctive shapes. The modern world was emerging: blame it all on wine.

✦ ✦ ✦

The Romans owed a huge cultural debt to the ancient Greeks, one that included their devotion to wine. By the time Rome had achieved hegemony around the Mediterranean after the Punic Wars in the third and second centuries B.C.E., the Romans found themselves at the center of the extensive wine trade that had originated in Canaan and Phoenicia and had subsequently been developed by the Greeks and Carthaginians. Indeed, the oldest Latin text that has come down to us, a detailed manual of farming practices written in about 160 B.C.E. by Cato the Elder, apparently leaned heavily on the work of the third-century B.C.E. Carthaginian Mago. This early agriculturist had also provided advice on every phase of winemaking from propagating, planting, fertilizing, irrigating, and pruning the vines, to grape pressing and fermentation. Mago's original Punic text has vanished; but Cato's instructions illustrate how sophisticated winemaking had become by his day, and how closely wine itself was integrated into the rapidly growing economies of Mediterranean countries.

Eventually, wine estates expanded so greatly that cereal production practically ceased on the Italian peninsula, making Rome dependent on its North African colonies for its grain supplies even as it exported increasing quantities of wine to the periphery of its empire, squeezing beer production in the process. As colonial peoples began to acquire a taste for wine, they began producing it locally. Although vine growing was prohibited beyond the Alps in 154 B.C.E. (to encourage exports), local viticulture (until the third century C.E. restricted to Roman citizens) gradually became established in what are now the classical northern European viticultural areas, especially France and Germany. Indeed, by the end of the first century B.C.E. French wines had gained a considerable reputation among Roman oenophiles. Thanks to the Carthaginians, Spain had by then long possessed a thriving vine-growing industry; Iberia helped make up the shortfall when Italian production mysteriously dipped in the second century C.E.

Especially after the Romans had discovered that burning sulfur candles inside empty wine jars would keep them free of vinegary smells, and so began adding sulfur dioxide as a preservative, wine became a durable product that could be taxed according to its quality. Much of the payment of such taxes was in kind, and this practice gave the Roman authorities reserves of wine to distribute, both to cement existing alliances and to buy off "barbarians" who might have threatened the imperial fringes. Over the centuries, for example, Rome sent large quantities of wine to Gaul, where humbler wines had been produced on a small scale ever since the Etruscans introduced wine to the region around 500 B.C.E. Imported Roman wine was shipped to ports on the Rhône estuary, where the local Celtic traders developed the habit of transferring it from amphorae into oak barrels before sending it on upriver to be traded for honey and timber. Thus was born one of the most hallowed regional winemaking traditions, as the new wine-storage technology underwrote the inexorable advancement of viticulture up the Rhône valley and into the French interior, even in the face of resistance from winemakers closer to Rome.

The best wines of the empire inevitably found their way to Rome, where they were prized as symbols of prestige and wealth. Everyone seems to have agreed on which these were, and at the pinnacle of repute were wines that came from the slopes of Mount Falernus, to the north of Naples. Made from the Aminean grape, these golden or amber-colored wines were probably high in alcoholic content, since Pliny the Elder recorded that they might "take light" when a flame was applied to them. The most fabled Falernian vintage was harvested in 121 B.C.E. Not only was it widely praised at the time, it was served to Julius Caesar a hundred years later, presumably to his entire satisfaction because someone was apparently brave enough to offer it again to Caligula in 39 C.E., when it was 160 years old.

✦ ✦ ✦

In the Greek and Roman traditions the consumption of liberal quantities of wine had been associated with the cults of Dionysus and Bacchus, gods of a pretty generalized hedonism. Still, while in both cases the value of wine lay in its practical role in the shedding of inhibitions, rather than with hierarchy and spiritual symbolism as in Egyptian tradition, it had great symbolic significance in the Hellenic and Roman worlds as a badge of civilization. But although the importance of wine to Rome itself was overwhelmingly social and economic, one side effect of Roman colonial activities was to diffuse the drinking of wine into peripheral regions where its consumption could be adapted to new contexts. And, as it happened, one of the unintended consequences of the road and sea transport systems established by the Romans to unite their empire was that not only was the transport of wine and other goods facilitated. So also was the spread of an obscure religion that had its origins in the Levant, an ancient winemaking region, at the beginning of the first century C.E.

The founder of that religion, Jesus Christ, grew up in a tradition that was steeped in wine, which his Jewish community considered a God-given blessing when consumed in moderation. Excessive inebriation was strongly disapproved of, and was condemned by biblical tradition to such an extent that some sects banned the consumption of wine. But the beverage was for the most part favorably viewed by Christ's community; after all, Noah's first act when he disembarked from the ark was to plant a vineyard. In Christ's time the average privileged citizen of his Judean homeland drank about a liter of wine a day; and, as recounted in John's Gospel, Jesus's first miracle involved saving an unfortunate situation at a wedding by turning six pots of water into reportedly excellent wine. Throughout the accounts of Christ's career, wine and vines crop up as recurrent themes: he likened himself to the vine and his apostles to its branches, and most significantly, during the Last Supper Matthew, Mark, and Luke all record that Christ gave his disciples wine, declaring, "This is my blood of the new testament." The offering of wine at a Passover Seder was hardly unusual:

the ceremonial drinking of wine was entrenched in Jewish tradition. But in light of Christ's remark wine took on a special significance for his followers, and thenceforward Christians imputed to it a specific symbolic role as the embodiment of the blood of Christ.

From its earliest days, the church celebrated the Eucharist with wine, and those in the mainstream disapproved of the Gnostics, who celebrated it with water. The practice meshed nicely with established habits that proved durable even as economic and political changes roiled the Levantine region. When, in the early fourth century, Constantine adopted Christianity as the official religion of the Roman Empire, his motivations were largely political (you will look in vain for any mention of Christ's teachings in the fourth-century Nicene Creed); but although its political aspect ultimately led to the church's bureaucratization, sacramental practices continued unaffected—as indeed they did following the division of the Roman Empire into eastern and western empires in 395. It is thus possible to discern a distinct continuity between Roman and Christian beliefs and imagery. Both Christ and Bacchus were thought to have been born of a god via a mortal woman, and both were associated with life after death. Bacchus had even previously used Christ's trick of turning wine into water, and scholars have found various other Bacchic symbols embedded in early Christian mythology. In symbolic as well as gustatory ways, wine formed a bridge between the ancient and the nascent modern world.

The five centuries after the sacking of Rome by the Visigoths in 410 and by the Vandals in 455 are often known as the Dark Ages. But although the city of Rome almost disappeared, and chaos ruled in some parts of its lost empire, many formerly colonial economies continued to flourish, or at least to muddle along. Most important, the vine-growing tradition persisted almost everywhere it had been established. Indeed, a taste for wine turned out to be one of the most durable aspects of Roman influence, waning only where climatic conditions were unsuitable for growing grapes.

Oenophilia was also a hallmark of many pagan tribes, but the symbolic role of wine as the blood of Christ accounted most strongly for its spread in Europe, a rapidly Christianizing region. There, in the face of a general decline in literacy, monasteries and other religious settlements often came to assume the role of guardians of historical, cultural, and agricultural knowledge. Initially most ecclesiastical establishments lacked the resources for more than limited wine production, restricting themselves to what was necessary for sacramental requirements and to maintain the quality of monastic life. But over time, some became famous for their wines and acquired increasing expanses of vineyard that were leased to local viticulturists, thereby helping revive the wine trade.

This persistence of the vinous tradition held true initially for all parts of the former Roman Empire. As in Europe, vines continued to be cultivated in such places as Mediterranean North Africa, the Levant, Persia, and even the Central Asian oases that dotted the fabled Silk Road, a web of trading routes extending toward China. But in the seventh century, with the rise of Islam, this long-established pattern was widely disrupted. From their Arabian place of origin, Islamic armies had by the middle of the eighth century conquered most of the Middle East and Mediterranean North Africa, as well as the Iberian Peninsula in Europe. And where Islam went, viticulture, or at least the making of wine, stopped.

The story goes that the young Prophet Muhammad had one day happened on a wedding at which wine was being consumed, and all the guests were happy and convivial. He left the feast murmuring blessings upon wine. But when he returned the next day, he found the place a wreck, the revelers bloody and battered from all-night drunken brawling, and revised his blessing into a curse. Thenceforward he forbade his followers to drink wine. In his view of paradise rivers flowed with this delectable liquid, but humans on earth could not be trusted to drink it without abuse.

There has been a lot of scriptural exegesis aimed at understanding precisely what Muhammad prohibited, and interpretations vary. One way in which wine production was stopped was the banning of the clay receptacles in which it was made. But skins were still permitted, and it is believed that Muhammad's own wives used wineskins to make him a potion prepared by soaking dates or raisins in water and allowing them to ferment slightly. The Arabic name for this concoction is *nabidh,* usually rendered in English as "date wine." But the accuracy of the translation is subjective and contentious, and in the Islamic world interpretation has increasingly tended toward a blanket ban on alcohol. From time to time, and place to place, a more relaxed take on the Qur'anic injunction has been adopted, and at the end of the eleventh century the Persian poet Omar Khayyam was still able to muse: "I often wonder what the vintner buys / Half as precious as the thing he sells." But in general, in most places where Islam imposed itself—and stayed—wine production and consumption ceased.

Still, it would be inaccurate to characterize the Islamic and Christian worlds as abstemious and bibulous, respectively . Even today, some Islamic countries take a softer stance on the matter of wine and other alcoholic beverages, while in the Christian world attitudes vary hugely, and the co-existence of the pleasures and pitfalls of wine have led to severe cognitive dissonance at both the individual and social levels. Perhaps the best example of this is provided by Prohibition in the United States of America.

✦ ✦ ✦

In the early days of the nation, Thomas Jefferson and some of his patrician colleagues were noted wine connoisseurs, reveling particularly in the wines of France. On a more populist level Benjamin Franklin wrote that "wine [is] a constant proof that God loves us, and loves to see us happy." But by the early nineteenth century, particularly owing to rapid urbanization, alcohol consumption and abuse had skyrocketed in the United States, leading to the development by around 1840 of a vociferous temperance movement. After the abolition of slavery many churches and secular associations also began to turn their abolitionist energies toward demon drink, first trying to persuade imbibers on an individual level to moderate their habits, and ultimately badgering state legislators to prohibit alcohol altogether.

By the end of the nineteenth century, the women's temperance movement in particular (women and their children tended to be the ultimate victims of men's excessive alcohol consumption) had achieved notable successes at local levels, not least as a result of the intensive media coverage of exploits like Carrie Nation's campaign of smashing up bars with an ax. On such victories was built the Anti-Saloon League, perhaps the best organized of all early lobbying organizations, which took aim directly, and effectively, at legislators' voting records. With its roots in conservative Protestantism and the explicit goal of countrywide prohibition of alcohol, the League rapidly managed to build a powerful coalition out of such unlikely bedfellows as the suffragists, the Ku Klux Klan, the Industrial Workers of the World, and John D. Rockefeller. A number of unlikely events then concatenated to move the League's agenda forward as the early years of the twentieth century progressed. An important element in its success was the domination of brewing in the United States by German immigrants, against whom anger could easily be whipped up as America entered World War I; indeed, drinking beer became downright unpatriotic. Also significant was the introduction, just before the war, of a federal income tax, which reduced the government's dependence on alcohol taxes. Importantly also, the argument for prohibition was couched mainly in moral terms, something that has always appealed to Americans. As a result, with their attention on other pressing issues, politicians were vulnerable to pressure by special interests to get prohibition legislation passed. By the end of 1917 the Eighteenth Amendment to the Constitution, banning the manufacture and sale of alcohol, had breezed through both houses of Congress, and, after rapid ratification by the states, it went into effect in early 1920.

Apparently, few of those who had supported the change had given much thought to its practical effects. Perhaps the only ironclad rule of human experience is the law of unintended consequences, and in the case of Prohibition this unwritten rule went into operation with a vengeance. Outlawing alcohol proved to have little if any effect on the demand for alcoholic drinks; the main result of prohibiting them was to increase prices and, as with today's war on drugs, to turn gangsters into millionaires. Other unanticipated economic effects included a general depression of economic activity and the impoverishment of local governments owing to the loss of liquor taxes. Ironically for a measure that was based to a great extent on moral outrage, the forbidding of alcohol to a population that still wanted to consume it had the paradoxical effect of causing widespread immorality in the form of flouting of the law. Almost everyone became a lawbreaker, and corruption was rife as many enforcement agents joined with the gangsters to profit from the booze business. Such anarchy could not last: the Eighteenth Amendment was repealed at the end of 1933, largely on the compelling grounds that it had led to a severe loss of respect for the rule of law.

Prohibition and its enabling laws present us with a prime example of well-meaning anti-alcohol legislation gone astray, but not with a unique one. Within the twentieth century alone, the sale of alcoholic beverages was banned for various periods in majority Christian countries as diverse as Russia, the Faroe Islands, parts of Scandinavia, and Hungary—always for the same stated reasons. For, while it is an unquestionable augmenter of the pleasures of life, this gift of the gods is also liable to hideous abuse, and has been responsible for the infliction of enormous misery. Viewed in this context, alcohol appears as a mirror for humanity itself. It is emblematic simultaneously of civilization and savagery, and it reveals the worst as well as the best in human nature. As a result, as long as alcohol produces its contradictory effects (which is to say, as long as our difficult and complex species remains in existence), human beings will continue to have a conflicted, contradictory, and complicated relationship with wine and other alcoholic drinks.

2

Why We Drink Wine

W hy do humans drink alcohol? Not implausibly, it has something to do with their primordial fruit-eating heritage: the scent of spontaneously fermented ethanol guided humankind's ancient arboreal ancestors to the ripest and most sugar-laden fruit in the tree. So we were pleased to obtain a bottle of inexpensive wine from New Zealand whose label made due obeisance to the primate that had inspired this "drunken monkey" hypothesis. True, the Sauvignon blanc grape and howler monkeys have no specific affinity, but the wine itself was amazingly drinkable, with all the extroverted grassiness and grapefruit finish one might expect from the grape and the place. We think the monkey would have approved.

At this point, it might be relevant to pause for a moment to consider why human beings are so fond of wine and other alcoholic beverages. Actually, humans are far from alone in their predilection, and, conveniently, naturally occurring ethanol, the kind of alcohol found in wine, occurs widely in the environment. Indeed, it is found anywhere plants produce sugars. Honey aside, the best concentrated source of sugars is fruit, and once the flowering plants began to diversify late in the age of dinosaurs, well over a hundred million years ago, fruit rapidly became available almost everywhere vegetation could grow. Large numbers of different animals began to specialize in fruit consumption, and at the same time other organisms began to colonize fruit as an environment for sustaining life. Prime among these latter were the yeasts, tiny single-celled fungi that we'll consider in detail in Chapters 5 and 6. Yeasts today implement fermentation of fruit sugars to produce ethanol, and it has been suggested that their ancestors began doing so as a way of discouraging other microorganisms from competing for the living space and sugars offered by the skins of oozing fruit. Because alcohol is toxic to many organisms, this explanation appears plausible; and in any event the natural fermentation of sugars by yeasts has become ubiquitous. Most of the time the concentrations of alcohol produced spontaneously by yeasts remain pretty low, but the phenomenon is widespread enough to help account for the fact that many different kinds of organism, particularly fruit-eating ones, possess the capacity to detoxify alcohol in small doses.

Some organisms, including humans, seem to benefit from consuming moderate quantities of alcohol. When scientists exposed fruit flies to vapors containing low, moderate, and high concentrations of ethanol, for example, the "moderate drinkers" lived longer and had more offspring than the abstainers. Why this was so is not clear, but it is well established that the scent of ethanol is an important factor in guiding flies toward sources of fruit, meaning that alcohol plays an important role in their economic lives—and in other aspects of their existences, too: fruit fly larvae plagued by parasites medicate themselves by seeking out ethanol-containing foods, and the scientists who observed this phenomenon suggested that alcohol might have a similarly protective effect in other organisms as well. Even if the alcohol did not save their lives, it might have at least cheered the afflicted flies up: another group of scientists reported in 1977 that male fruit flies deprived of the opportunity to mate showed a stronger preference for ethanol than their more successful counterparts.

The key here is quantity. In fruit flies, large quantities of alcohol negate the benefits of small ones, illustrating a common phenomenon known as hormesis, whereby substances that are toxic to animals in large doses can have favorable or agreeable effects in small ones. Hormesis is widespread in nature; and although scientists still debate how it works, one idea is that low levels of many toxins activate physiological repair mechanisms in the body that have broader effects than simply responding to the toxin. Another suggestion is that toxins in low concentration may promote antioxidant effects in the body. Whatever the case, alcohol does seem to have certain favorable effects on humans when taken in moderation, as attested by a substantial literature on the health benefits of drinking wine.

Despite such intriguing discoveries, though, how widely any health benefits of alcohol consumption are enjoyed in the animal world is not known. It is clear that some mammals simply like the stuff, which is hardly surprising, because elevated blood-alcohol levels appear to enhance the production of epinephrine (adrenaline), a hormone that acts in the brain to reduce inhibitions. Thus elephants in parts of southern Africa have long been famous for excessively indulging in the naturally fermenting fruit of the amarula tree, weaving away rather unsteadily after each episode. But as with most other mammals that occasionally go in for such overripe fruit while it is still on or around the tree, this is a strictly seasonal predilection for elephants, limited to the short periods of the year during which the alcoholic fruit is available. Heavy "drinking" is hardly a way of life for them.

More remarkable, then, is the case of the tiny pen-tailed tree shrews of Malaysia. They are particularly instructive in a consideration of why human beings like alcohol because they are widely reckoned to be among the closest living relatives of the primates, the zoological group to which our species *Homo sapiens* belongs. They may not be precisely the same as our ancient ancestors from the beginning of the age of mammals some sixty-five million years ago, but in appearance, body size, and general habits they probably come close.

In 2008 German researchers reported the results of an ecological study of pen-tailed tree shrews in a western Malaysian rainforest. They noted that the little creatures, under 50 grams in weight, returned repeatedly to feed on the large flowers of the trunkless bertam palm, an abundant plant of the forest floor. Throughout the year, for prolonged periods of time, these flowers exude nectar to attract pollinators. Frothing and bubbling, along with "brewery-like" odors, indicate that the nectar is colonized by natural yeasts and begins to ferment virtually as soon as it is produced. The resulting alcohol concentration in the nectar is as high as 3.8 percent, about as strong as most of the beer traditionally sold in the United States. During the course of the study several species of mammal visited the bertam palms each night in search of this resource, including our primate relative the slow loris; but the pen-tailed tree-shrew

Pen-tailed tree shrew at the bertam palm bar

beat them all in its enthusiasm for the nectar, sometimes spending well over two hours per night bingeing on this delicacy. Oddest of all, the tiny creatures never seemed to get drunk—even when, relative to their body size, they had imbibed enough to cause a large man to pass out from inebriation. Elevated alcohol levels were detectable in the tree shrews' blood, but the tiny mammals showed no physiological impairment—fortunately, since these vulnerable creatures are under constant threat of predation. Their senses have to be keenly attuned to danger at all times, and their reactions have to be swift. It is clear, purely from the time spent feeding, that the palm nectar is an important nutritional resource for the tree shrews at the Malaysian study site—as indeed beer, which is actually more nutritious than bread, can sometimes be for humans. But if the tiny quaffers had lacked appropriate mechanisms for counteracting the physiological effects that alcohol has in us, they would have been in deep trouble.

The example of the tree shrews suggests, at the least, that from the beginning of primate history there may have been both an occasional predilection for the products of fermentation and a mechanism for processing alcohol. Humans exhibit this primate heritage in their physiology: by some reckonings, about a tenth of the human liver's processing capacity is slanted toward breaking down alcohol via production of such enzymes as alcohol dehydrogenases. Scientists often marvel at this huge apparent dedication of resources to one specialized task, although in fact the alcohol breakdown apparatus benefits from a fortuitous resemblance between ethanol and other molecules which are more routinely encountered.

There are also differences of scale at work here. Although a tiny tree shrew can slake its thirst for alcohol by licking palm-tree flowers, this is hardly a viable solution for human beings, with well over a thousand times more body mass. Naturally occurring sources of alcohol hardly fill the bill, which is why it is more likely that, from the time when they first gained the intellectual and technological wherewithal to figure out how to do it, humans have been devoted to the artificial production of alcoholic beverages.

It is of course a long way, evolutionarily speaking, from tree shrews to *Homo sapiens*; but we have larger and closer relatives that also share our alcoholic proclivities. Howler monkeys in Central America are much bigger than tree shrews (they can weigh up to 9 kilograms) and have been seen to feed frenetically on the gaudy orange fruits of the *Astrocaryum* palm. In the best-known case, the sheer exuberant enthusiasm of a particular howler in a forest in Panama aroused scientists' suspicions that it might be drunk. These suspicions were quickly confirmed. Combined with the observed quantity of the orange delicacies consumed, analysis of alcohol in partially eaten fruit the monkey had let fall from the tree showed that he had consumed the equivalent of ten bar drinks in a single session. No wonder he was tipsy by the end of his feast! And although the scientists observing him reported no immediate adverse physical consequences—he did not fall out of the tree, at least—they did not check him for a hangover.

Observation of the happy howler fit in neatly with the "drunken monkey" hypothesis that the biologist Robert Dudley developed early in the twenty-first century as an evolutionary explanation of the human predilection for alcohol. Dudley pointed out that our heritage is a fruit-eating one. Almost certainly, the first primates were frugivores, and although some of our early relatives soon moved on to leaves and other nonfruit plant parts, the hominoid (ape/human) group from which we emerged some seven million years ago had clearly stayed on the fruit-eating path. Ethanol "plumes" emanating from fruit can be as useful to keen-nosed primates for locating sources of food as they are to fruit flies, and Dudley suggested that early fruit-eating monkeys and hominoids were attracted to ripe fruit by alcoholic aromas.

The plausibility of this scenario was reinforced by the demonstration in 2004 that the presence of ethanol is a better indicator of how much sugar a particular piece of fruit contains than its color is. Even more important, once they have started eating fermenting fruit, the feeders are rewarded by a further energetic premium: unusually for an addictive substance, the caloric value of ethanol is high. Its calorie count is, in fact, almost double that of carbohydrates, as every beer belly in the world attests (if there were no beer around, doubtless we'd be calling them wine bellies). So the idea here is that a variety of circumstances would have conspired to make alcohol attractive to our hungry fruit-eating ancestors, and this predilection was passed down to their descendants today.

The "evolutionary hangover" hypothesis is attractive, but it has some difficulties when applied specifically to humans. A special aspect of the evolution of our own African ancestors was that when they began to venture out of the forests and into woodland and tree savanna environments several million years ago, they changed their diet significantly. Chimpanzees roaming through comparably open areas today largely ignore the new sources of sustenance potentially available to them, and stick to a diet mainly consisting of fruit and leaves, the forest resources with which they are familiar. In contrast, our own hominid ancestors became the ultimate omnivores, reducing their intake of fruit and adding foods such as bulbs, tubers, and animal proteins to their menu. Our ancestral break with the forests thus involved abandoning fruit as the dietary mainstay. Additionally, in sharp contrast to that Panamanian howler, even in the forests many monkeys and apes have been observed actively avoiding the overripe fruits in which ethanol concentrations are highest. As a result, it is impossible to generalize about whether frugivorous higher primates as a whole like alcohol. Some do, and they evidently enjoy not only the valuable information it gives them about fruit quality, but also its behavioral effects.

From a human perspective, one of the most important implications of our descent from an ultimately fruit-eating stock is that our ancient ancestors would inevitably have been routinely exposed to a low alcohol concentration in their diet, whether they actually sought it or not. Such sustained—if muted—ancestral exposure may partly explain how modern humans have come by their modest physiological ability to detoxify alcohol, although a fortunate molecular coincidence is also involved. There is, of course, a limit to how much alcohol various animals can handle, and gram for gram tree shrews have an unusually high tolerance. But humans are remarkably unlike the hedgehog that is reported to have expired from drinking egg liqueur when the alcohol concentration in its blood had not reached even half the legal limit at the time for driving in New York State, and molecular scientists now think they know why. Apparently, a tiny DNA change in the last common ancestor of modern apes and humans resulted in the production of an enzyme that is super-efficient in breaking down the ethanol molecule. In light of this finding, it is perhaps less surprising that humans are attracted to ethanol than that apes do not more actively seek out fermenting fruit. But in any event, once *Homo sapiens* had acquired its creative bent, this unusual new genetic propensity gave our species an edge in employing fermentation as an economic tool.

✦ ✦ ✦

Ever since people took up settled existences and could no longer follow animals or plants around the landscape to where they were grazing or producing fruit or seeds at different times of year, they have faced the problem of how to store perishable food. Even with the best agricultural practices, a single fixed location will almost never be equally productive at all seasons. But keeping food around is hardly simple. Stored food rapidly rots through oxidation and other chemical processes, and it is also subject to the depredations of such pests as insects and rodents.

All settled humans thus require means of actively preserving food, and fermentation was probably widely recruited for this purpose by Neolithic people. The zoologist Douglas Levey has pointed out that from an anthropological point of view deliberate fermentation can best be seen as "controlled spoilage." Most microbes responsible for the decomposition of food cannot persist even in moderate concentrations of alcohol, so by permitting limited alcohol production in stored foodstuffs via controlled exposure to oxygen, Neolithic farmers were able to preserve much of the nutritional value of their crops, even if not their freshness.

For the record, though, fermentation is not the first form of food preservation documented. In the latest part of the last Ice Age, about fourteen thousand years ago, ingenious inhabitants of the icy Central European Plain were already storing meat in deep pits dug in the permafrost to create year-round refrigerators. In the balmy Neolithic Near East this technology was obviously not an option, although drying foods in the sun was doubtless another major approach to the preservation of foodstuffs, and would have provided an obvious solution in many cases. Still, fermentation was clearly important enough to Neolithic farmers as a food-preservation strategy for Levey to suggest that the process was initially adopted for this purpose, to be used only later in producing alcoholic beverages.

Members of our symbolically reasoning species mentally process information about themselves and about the world around them in a novel way. The results are remarkable. But we are unperfected creatures nonetheless; and behaviorally we are still bound by what statisticians call the normal distribution. Otherwise known as the bell curve, the normal distribution simply acknowledges that most people are broadly similar in behavioral and physiological expressions, and deviations from the average become increasingly rare toward the extremes. Most people behave reasonably decently toward one another, for example, while both the saintly and the monstrous are fortunately few. The same goes for the spectrum that lies between being teetotal and a heavy drinker, which explains why the abstemious and the alcoholic make up only minorities of the population. What is more, among humans both unwholesome social pressures and principled beliefs tend to exaggerate behavioral tendencies, as witness Saturday-night binge drinkers on the one hand, and temperance campaigners on the other. But the underlying pattern remains basically constant, and a quick look at the natural world makes it obvious that *Homo sapiens* is not the only species with a moderate tolerance for alcohol that also overindulges occasionally.

The big difference in our case, of course, is that *Homo sapiens* has devised means of producing abundant alcohol at will. Combine this ability with alcohol's addictive and disinhibiting properties, and it seems inevitable not only that some individuals will overconsume but also that this behavior will be recognized as a social evil. Virtually every human society has consequently produced strict rules to govern alcohol consumption, and given our species' predilection for taking any good idea to its illogical extreme, such rules have not infrequently been compulsively developed into rituals. Innumerable laws, traditions, and proscriptions may regulate the production, distribution, and consumption of alcohol. Yet at the same time, attitudes toward wine can vary within the same culture and even the same individual (humans are famously cognitively dissonant), ranging from seeing it as "the Devil's potion" to the "blood of Christ." This is why it is so easy to believe that the products of the winery at Areni-1 were both produced and consumed according to strictly conventionalized and rather compulsive procedures. And why conflicted attitudes toward wine and other alcoholic beverages have reigned ever since.

3

Wine Is Stardust

Grapes and Chemistry

The story of wine starts with stardust. And on any matter that embraces both stardust and wine we trust only one source: our oenophile friend and colleague, the astrophysicist Neil deGrasse Tyson. We asked Neil to choose a wine with an astrophysical name and theme, and he immediately came up with Astralis, the designation given by Australia's Clarendon Hills vineyards to its flagship Syrah. The grapes come from vines that look as old as the universe: so ancient, huge, and gnarly that they grow individually like trees, without any trellising. And the wine itself? We asked Neil. "Big," he said. "Bold. Beautiful. Radiant to the senses. Just like the stars themselves."

How do crushed grapes turn into wine? To explain this process we must go back to the very beginning—to atoms and molecules, the building blocks of both wine and the universe. The first and simplest atom, hydrogen, was formed from the strewn stardust that eventually made up the galaxies, stars, and planets. Thereafter, other elements began to combine, and the scene was set for our modern universe to evolve with all it contains. Neil deGrasse Tyson once declared, with his usual eloquence, that we are all, both figuratively and literally, made of stardust. And if this is true of humans, then it is true of wine as well.

A great metaphor for the cosmic origin of wine showed up some years ago when the astrophysicist Benjamin Zuckerman and collaborators discovered that a dense molecular cloud lying near the center of the Milky Way galaxy, to which our solar system belongs, contains alcohol. Hailed by

Tyson in a *Natural History* article as the "Milky Way Bar," it would actually be a bit of a disappointment as an earthly drinking establishment because the water molecules in the cloud vastly outnumber the alcohol molecules. In fact, as Tyson points out, in combination they would yield a libation of only about 0.001 proof. But out in the galactic vastness the cloud itself is so huge that, with sufficient distillation, its alcohol molecules could provide something "on the order of 100 octillion liters of 200-proof hooch."

Nearly every culture recorded has figured out a way to turn sugary concoctions into alcohol. Early brewers and winemakers had no knowledge of atoms and molecules, acids and bases, hydrogen bonds and electron orbits. Yet they were expert protochemists, controlling and tweaking one of the simplest, yet most important chemical reactions to human existence—the transformation of sugar into alcohol. Although our ancestors achieved the basics of the chemistry through trial and error, without understanding what was happening at the molecular level, today such an understanding can help the wine drinker appreciate why a wine is bitter or acidic, why the alcohol content of wine rarely exceeds 15 percent, why wine takes as long as it does to ferment, and why it should be stored with care.

Let's start with the scale of the molecular realm in which the chemical reactions of fermentation occur, which are vastly different from those of the Milky Way. Atoms such as hydrogen and small molecules such as water and sugar are minuscule. The glass from which you are drinking your Chianti might be about 10 centimeters tall and 5 centimeters in diameter, and weigh about 30 grams. In contrast, a typical atom, the basic unit of which molecules are composed, is between 25 and 200 picometers (pm) in size. (A picometer is .000000000001 meters, and is thus about 0.0000000001 times smaller than the height of your wine glass.) If you stacked these atoms atop one another it would take 100 billion of them to reach the rim! Water molecules are 2.5 angstroms (Å) in diameter. (An angstrom, another measure of size, is 0.00000001 meters.) So a water molecule is about 0.0000005 the diameter of a glass of wine, which is some 50 million molecules wide. A molecule of sugar, the major component of grape mash, weighs 180 times 0.00000000000000000001 grams. So a single molecule of sugar composes about 0.00000000000000000002 of the weight of the wine in your glass. In other words, you would need a million trillion molecules of sugar to achieve the weight of the wine in your glass. Fermentation involves the interaction of molecules, but the scale at which those interactions occur is obviously too small for the human mind to visualize directly, and the conversion of a mere gram of sugar into alcohol requires that trillions of reactions happen in a *very* tiny space.

✦ ✦ ✦

It is important to understand the atomic structures of alcohol, sugar, and other molecules because their compositions and shapes determine their natures and those of the interactions among them. Scientists have by now a sophisticated picture of what atoms are, but for the purpose of understanding wine we can imagine them as simple orbital structures consisting of a nucleus of clustered protons and neutrons with electrons spinning around the periphery. The orbits of the electrons around the nucleus can be really complex, and they make the physics of atoms pretty weird. But here we need only note what happens when a stable atom—that is, an atom with the same number of electrons as protons—loses or gains an electron. This loss or gain happens continually; indeed, if it didn't, probably nothing in the universe would have become more complex than an atom, and wine—and wine drinkers—would not exist. Electrons are able to do their in-and-out atomic hokey-pokey because nothing is protecting them from outside forces, or even from their own eccentricity, as they orbit around the nucleus.

When a lone electron joins with a stable atom, it produces an imbalance in the number of protons and electrons. Specifically, there will be more electrons than protons, and the atom will be negatively charged (have a charge of −1). Conversely, when a lone electron skips out of its orbit, the resulting atom is positively charged (has a charge of +1). The ultimate accountant here is the universe, which likes to keep the ledger in balance. Of course, any individual atom that has lost an electron can capture a new one, just as any electron-heavy atom can kick one out. But often the stability the universe desires is achieved by combining two different kinds of atoms with opposite charges. This interaction is the currency of higher molecular structure, and it is known as a chemical bond. In the case of a lost or gained electron, we have an "ionic bond," but other kinds of chemical bonds also exist, and are important in combining smaller molecules to form larger and more biologically important structures such as DNA and proteins.

Fortunately we can simplify at this point because, of the 115 elements in the periodic table, only a few are relevant to the biology of wine. And of those, only a handful of larger molecules are present in the wine we drink, because living creatures incorporate a limited number of elements. Indeed, animals contain only six major elements: carbon (C), hydrogen (H), nitrogen (N), oxygen (O), phosphorus (P), and sulfur (S). Most students nowadays remember this by using the mnemonic CHNOPS, although a more precise alternative would be OCHNPS, which lists the elements in the order of their abundance in animal bodies.

For the yeasts used in winemaking, the appropriate order would be the even clumsier OCHNClPS. This is because, although the elemental makeup of yeasts is about 99.9 percent the same as that of animals, yeasts also contain chlorine (Cl)—in relatively high abundance. For plants such as grapevines the situation is more complicated still, and the mnemonic (as it were) would be OCHNKSiCaMgPS. The new elements in plants are silicon (Si), calcium (Ca), magnesium (Mg), and potassium (K). This string of abbreviations may not be easy to memorize, but it is important because there are several more basic elements in plants than there are in animals, or even in yeast. Still, humans, yeasts, and plants all have the first four elements (OCHN) in common, and P and S are also in there somewhere. Significantly, OCHNPS are the basic constituents of amino acids, the molecules that make up proteins, and of the bases that make up DNA, to which we'll return in a minute.

Why humans are made up of OCHNPS, and not six other atoms, can be answered in one word—evolution. Early in the life of our planet, natural selection honed the interactions and shapes of molecules. At the start, the evolution of life could have gone in any of several directions. Molecules, for instance, have handedness—they coil either to the left or to the right. In the illustration depicting two amino acid molecules with the same atomic makeup, each has the same number of carbons, hydrogens, oxygens, and nitrogens, yet the two will behave differently because one is left-handed and the other right-handed. In your imagination, try to turn the right-handed one around to get the left-handed one. You can't do it; if you tried to get a chemical reaction to work with right-handed molecules where left-handed ones should be used, you'd fail. Most of the molecules important to life on earth have evolved to be left-handed, for no better

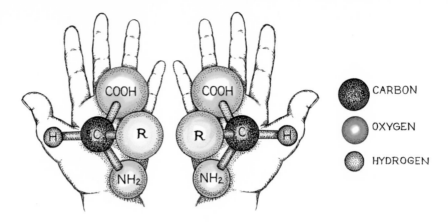

Typical structure of an amino acid (where R stands for any side group of the 20 amino acids), showing both the left-handed and the right-handed version

CARBON

OXYGEN

HYDROGEN

reason than that events early in the evolution of molecules dictated a general trend for all the molecules. And so it is with our six atoms and alcohol.

Energy is an important commodity for cells, and fermentation, the process that makes alcohols, is a fairly efficient way to produce it. In primitive cells, the need for energy probably dictated the honing of the fermentation process. But why fermentation persisted and won out in these primitive cells was more or less a matter of chance. In brief, evolution is a master tinkerer. So if the very primitive process of fermentation had ever been supplanted by a better mechanism, we might not have alcohol at all.

There are many ways in which the various elements we've been discussing can bind to one another, and it is this diversity that makes our world so complex, for it is the shape and spatial orientation of a molecule that largely dictates its behavior under various circumstances.

Let's take a closer look at one of the most basic of those elements: oxygen. This is the most abundant element on the surface of our planet, and it is incorporated into the bodies of all of earth's living beings because they are almost entirely made up of water, which contains oxygen in a ratio of 1:2 with hydrogen. Another major molecule in the atmosphere is carbon dioxide, in which oxygen is found in a ratio of 2:1 with carbon. Water forms as a stable combination of the three atoms that constitute each of its molecules: two hydrogen atoms and one of oxygen. The bond here is not the ionic type that entails charged particles, but rather involves making "mutual loans" of electrons.

In *Oxygen: The Molecule That Made the World*, Nick Lane emphasizes that life on Earth is dependent on two basic processes involving oxygen— namely, respiration and photosynthesis. It is not too much to claim that the entire "economy" of life on our planet is based on how these two processes move electrons around. Photosynthesis, in which organisms take up carbon dioxide and water and convert them to oxygen and energy, is unique to plants, algae, and some very small organisms such as the Cyanobacteria. Respiration, which sustains life in all living organisms, involves converting atmospheric oxygen to energy, water, and carbon dioxide.

✦ ✦ ✦

Chemists love equations, and to understand fully what makes up wine it is necessary at least to tolerate them. Deciphering a chemical equation might seem a bit like reading the Rosetta Stone, but with a few rules in hand it becomes quite simple. One way to write a chemical equation describing a molecule is simply to list the symbol of each atom in the molecule, with a subscript giving the number of times it occurs. So carbon dioxide, which has one carbon and two oxygens, is written as CO_2. But although this way of describing molecules tells us what atoms are present, it does not indicate how they are arranged or the shape of the molecule. And knowing this is critical to understanding a molecule's function. To add spatial information, chemists use "stick notation," which resembles the stick figures used in the game Hangman. Each atom has a typical number of sticks protruding from it. Thus hydrogen (for the most part) has only a single stick, while oxygen typically has two sticks, and carbon has four. The number of sticks protruding from a particular atom is dictated both by its atomic number and by the orbits of its electrons, and in stick notation carbon dioxide looks like this:

$$O=C=O$$

But the page on which this stick diagram is drawn is flat, and molecules exist in space, with a three-dimensional structure. So we need to discriminate between carbon dioxide's bookkeeping format above, and its natural (stick-and-ball) form as illustrated in the figure. In this case, the three-dimensional structure of carbon dioxide is similar to its bookkeep-

CARBON

OXYGEN

The natural (stick-and-ball) structure of carbon dioxide

ing structure. But in many other molecules the angles at which atoms are connected to one another are not as clear. This is important to biologists because, at the molecular level where fermentation occurs, nature likes shapes. It doesn't necessarily care what makes up the shapes, taking its cues from the molecules' external form. So now, with these tools of scale and chemical equations in hand, let's follow how a carbon atom in a grape is transformed into alcohol.

So far it's been pretty straightforward, but atoms actually bind at different angles to one another, and representing carbon dioxide on a single line does not mean that the bonds made between carbon and oxygen in the physical molecule lie on a 180-degree continuum—a distinction that obviously affects their shape. Molecules of different shapes behave differently. This is a recurring theme in chemistry and biology, and when we discuss proteins (larger molecules with specific cellular functions) later in this chapter, we will see that altering the shape of a protein even slightly changes its behavior. In extreme cases it will challenge the viability of the organism that produces it.

Of all the many molecules that make up wine, the alcohol molecules are perhaps the simplest. There are several kinds, all conforming to the ball-and-stick structure in the figure, in which the red and light-gray balls represent oxygen and hydrogen, respectively: the dark-gray ball is carbon,

CARBON

OXYGEN

HYDROGEN

Stick-and-ball structure of a generalized alcohol molecule

and the Rs are simple side chains made up of carbons and hydrogens. In the saturated state, the central carbon should be fully bound to other atoms, meaning that there are three side chains, or groups, sticking out from it. The fourth, the OH sticking out from the central carbon, is called the hydroxyl functional group, and is found in all types of alcohol.

The smallest alcohol is methanol, in which $R^1=R^2=R^3=$Hydrogen (H). Simply envision each of the Rs in the illustration as an H, and there you have methanol, which has the chemical equation CH_3OH. It can be obtained by distilling wood—hence its alternative name of "wood alcohol"—and is pretty easy to make, notably as a byproduct of poor distilling procedures. The desirable alcohol that enlivens wine and beer and other alcoholic beverages is called ethanol, and it is a molecule in which $R^2=CH_3$ and $R^1=R^3=H$, all connected to the central C in the diagram. Hence its atomic equation is C_2H_5OH. This lovely molecule is the one we are after in winemaking and brewing, but it is only subtly different from the poisonous methanol molecule. What makes the difference is the simple CH_3 group connected to the central carbon. The tiny distinction between the harmful methanol:

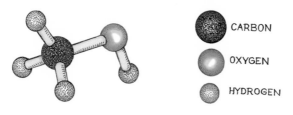

CARBON

OXYGEN

HYDROGEN

Stick-and-ball structure of methanol

and the benign ethanol:

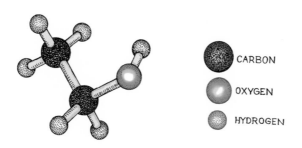

CARBON

OXYGEN

HYDROGEN

Stick-and-ball structure of ethanol

makes all the difference between becoming extremely sick (and potentially blind—methanol has a special hatred for the optic nerve) and being pleasantly tipsy.

Two other kinds of alcohol are important, too, because they are by-products of fermentation by bacteria and yeast. These are butanol and propanol, molecules produced during fermentation by, respectively, a bacterium named *Clostridium acetobutylicum* and otherwise innocuous yeasts at high temperatures. Both molecules are unwanted contaminants of beer and wine: the ethanol comes from the breakdown of sugars, and the methanol, butanol, and propanol from the breakdown of cellulose.

The desirable ethanol is a simple molecule, merely a couple of carbons, a handful of hydrogens, and an oxygen. But how those atoms are arranged and the space they fill are critical to how an alcohol molecule affects the nervous system. Simply by dropping the CH_3 from ethanol we literally produce a killer alcohol.

Now let's look at the sugars, molecules that are critical in the winemaking process. Like alcohol, they come in various guises. The most familiar sugar is sucrose, the one we use to sweeten coffee. Along with its cousins maltose and lactose, sucrose is a disaccharide sugar, a combination of two monosaccharide sugars, such as the common fructose and glucose. (More complex combinations are called polysaccharides.) The basic monosaccharide sugars form through glycolytic chemical bonds. What is important to note is that, when a glycolytic bond is formed between two monosaccharides, water is released. Such bonds are extremely strong, and can be weakened only by hydrolysis, the process of bringing water back in.

The molecular structures of sugars take ring shapes, in contrast to the linear molecules of alcohol and water. Some monosaccharide sugars take up a pentose or hexose form, according to how many "points" (five or six) there are in the ring. Carbons lie at the corners of each molecular ring, and different groups stick up and down from them to balance the chemistry. Sugars seem sweet to us because the atoms hanging off the ring interact with the taste receptors on our tongues. This is also why different sugars have different sweet tastes, since different shapes of sugar molecules make different kinds of contact with our taste receptors. What dis-

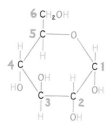

Chemical structure of glucose

tinguishes one hexose sugar from another is the nature of the side groups sticking out from the basic stop-sign-shaped structure. Consider, for example, the sugar ring for glucose: the carbons in the ring and in the side groups can be numbered like the numerals on a clock face. In the form of glucose illustrated here there are six carbons, and we can number them starting at three o'clock. Note that the hydroxyl groups (OH and HO) either stick up or down. The order of these OH groups is important in defining the overall structure and shape of sugars; most important, it determines how the molecule behaves. In this glucose molecule, the order of the OH groups from carbon positions 1 to 4 is down, down, up, down.

But by flipping the OH group in the number 2 carbon, so that it is on the upside of the ring, we can make a different sugar: a form of mannose which, while sweet, is unstable and not found in nature: in this case the order of the OH groups, from carbon 1 to carbon 4, is down, up, up, down. The difference counts: yet another form of mannose, in which the OH

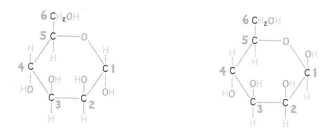

(*left*) Chemical structure of unstable mannose

(*right*) Chemical structure of bitter mannose

groups attached to the 1 and 2 carbons of glucose are both flipped, to give up, up, up, down, actually has a bitter taste. Predictably, there are precisely sixteen permutations for the positions of OH groups.

✦ ✦ ✦

What powers living systems on earth is the energy of the sun, which is essential for plants (and in turn, because animals eat plants and other plant eaters, for us animals, too). To make the energy their cells use, plants capture sunlight, and the hallmarks of fruits such as the grape is that they are packed with polymeric sugars derived from photosynthesis, the chemical reactions that occur in the chloroplasts of plant cells. Plants acquired these organelles in the remote past through the cannibalistic engulfment of a bacterium. The photosynthesis chloroplasts make possible is crucial to producing the sugar molecules that are such an important component of wine.

Photosynthetic cells in plants depend on various small molecules, the most abundant of which is chlorophyll, the pigment that gives leaves their green color. Chlorophyll absorbs light very efficiently, but only in the red and blue ranges of the color spectrum. Because chlorophyll does not absorb light in the green range it is reflected, which is why we perceive the leaves as green (for more on how we see colors, see Chapter 9). The chlorophyll molecules are packed into a region of the chloroplast called the thylakoid membrane, where they capture energy and transfer it to other chlorophyll molecules. From the point of view of wine, the most important aspect of photosynthesis is that sugars are produced as a by-product of this energy transfer.

Plants have a second way of storing the energy produced by photosynthesis: namely, by removing electrons from substances such as water. The loose electrons are used to make carbon dioxide and convert it into larger carbon-containing compounds, such as sugars, that are great sources of energy. The most important of these energy sources is glucose, and by making long chains of linked glucose molecules plants can store energy very efficiently. The resulting long-chain molecules may be of various kinds, including starch and cellulose. Neither tastes sweet, because both molecules are too big to fit into the taste receptors in our mouths.

Starch is made up of two kinds of molecule. One is amylose, a simple straight-chain molecule in which glycosidic bonds connect the glucoses to one another. The second, amylopectin, while partly linear, also branches. The powdery substance we recognize as starch once it has been removed from plant cells is about three parts amylopectin and one part amylose. In contrast, cellulose is composed of glucose chains that are also linked by glycosidic bonds, but come together to form sometimes structurally rigid lattices. Paper is made of cellulose, which is also a major component of such foods as lettuce (we are exhorted to include lettuce and other leafy green vegetables as roughage in our diets because the cellulose is barely broken down by our digestive tracts). Significantly, although celluloses and starches are both made of long chains of glucose molecules, they behave quite differently. Grapes contain both starch and cellulose, and hence a large amount of glucose as well as fructose. Both sugars come ultimately from sucrose that is produced by photosynthesis in the leaves of the grapevine, and has been converted to fructose and glucose by the time the sugars reach the grape.

The production of sugar in plants, and in grapes especially, does not occur spontaneously. Cells contain larger molecules called proteins. These act as machines, doing various jobs around the cell. Grapes are full of them (as are all living things), and grapevines continuously churn out the proteins that are essential for cellular function. At the same time, single-celled organisms such as yeasts are also constantly manufacturing proteins to maintain their internal housekeeping and deal with environmental challenges. Proteins are made of simple building blocks called amino acids, and amino acids have a basic core structure much like that of the

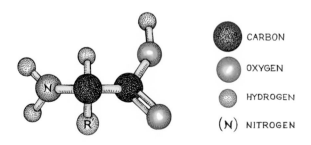

Generalized stick and ball structure of an amino acid

sugars described above. The general stick-and-ball structure of an amino acid can be seen in the figure. Note that there are two ends to an amino acid: an amino end (H_2N) and a carboxylic end (COOH). In the middle lies what is called a central carbon, and off this come a hydrogen atom and a chemical group called R′. The notation R′ stands for a side group, any one of a group of about twenty (sometimes more) chemical structures that may be placed in this position. The identity of the side group dictates the chemical, biological, and physical properties of the amino acid.

The simplest amino acid (in terms of number of atoms) is glycine, in which R′ is a single hydrogen (H). Having an H in this position gives the amino acid polarity, but it is balanced in charge, or what is called "uncharged." Substituting a methyl group (CH_3) for the R′ results in the amino acid alanine. This is charged, and thus both hydrophobic (water-repelling) and nonpolar. This tiny change gives alanine chemical behaviors that differ from those of glycine. The heaviest amino acid is tryptophan, which has an enormous number of carbon, hydrogen, and oxygen atoms in its side group. The side group has little impact on polarity and charge, but it is so bulky that it powerfully affects the shape of the proteins in which it occurs.

Just like sugars, proteins tend to make long chains. They are obliged to, since the carboxyl group on one end of an amino acid is attracted to the amino group on the end of another. Preference for this reaction makes proteins resemble beads on a necklace. And, as anyone who has ever had to untangle headphone wires knows, a linear array can fold and roll into a shape that is far from linear, is sometimes rigid, and is tough to unravel. But, unlike the headphone cord, whose tangles are random, proteins fold based on the arrangement of the primary beads on the string (twenty kinds, in the case of amino acids). Different orderings of amino acids along protein chains produce differences in how they fold to assume three-dimensional shape, and give them a wide array of shapes and hence functions.

Scientists often describe proteins or enzymes as molecular machines. Some are stand-alone machines, individualists able to do their job without assistance. Other proteins are like parts of an old grandfather clock, among which intricate interactions are required for proper functioning.

Many of the molecular machines relevant to winemaking and alcohol production are loners that add a phosphate here or split a bond there. But all form part of linked chains of reactions that have been honed by nature over millions of years.

Although many plant proteins are important in producing sugars, pigments, and other molecules crucial to winemaking, by far the most important are those yeast proteins that participate in fermentation. This conversion of sugar into alcohol is the product of three subprocesses carried out by two complex molecular machines and one simple chemical reaction. The goal of the first machine is to make a small molecule called pyruvate out of larger sugars such as glucose. The second machine then converts pyruvate into a smaller molecule called acetaldehyde. Finally, a simple chemical reaction converts acetaldehyde into alcohol. The first machine is a complicated one, involving several proteins linked together into a larger machine that carries out glycolysis. Following a specific carbon through glycolysis requires a knowledge of all nine of the protein submachines involved, and of the functions of those machines—which is mostly to add something like a phosphate (P) to the reacting molecule or to break a bond. In addition, electrons are moved around by another molecule called nico-

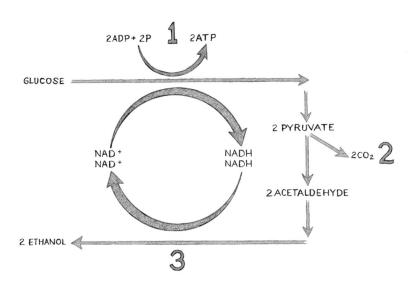

Converting sugar into ethanol

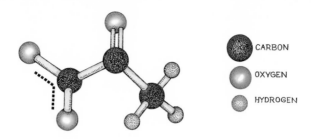

CARBON

OXYGEN

HYDROGEN

Stick-and-ball structure of pyruvate

tinamide adenine dinucleotide phosphate-oxidase (NADPH, which helps produce NAD+ and NADH, as will be described below). In this book, we will not dwell on the details but note only that the machinery involved is so exquisite that some proponents of Intelligent Design have used glycolysis as an example of irreducible complexity. So let us hasten to add that thinking about glycolysis in this way is hugely misleading because the steps of glycolysis mimic the evolutionary process by which eyes are thought to have evolved—namely, through a series of common ancestors in which intermediates existed.

Pyruvate is a small molecule that contains three carbons, three oxygens, and three hydrogens: The dotted line on the left in the stick-and-ball figure indicates that the two oxygens bound to the carbon at the vertex where they are connected share an electron, an arrangement that makes pyruvate very reactive. In making alcohol, the machine that breaks pyruvate down is called a decarboxylase because it removes a carboxy group.

The machine takes in the reactive pyruvate, removes the carboxy group on its right, and releases acetaldehyde, as seen in its stick-and-

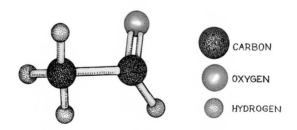

CARBON

OXYGEN

HYDROGEN

Stick-and-ball structure of acetaldehyde

WINE IS STARDUST

ball diagram. Remember that nature is a strict bookkeeper, so the "before" number of atoms must be balanced by an added hydrogen in the "after" version. When the carboxy group on the far right is replaced by a hydrogen, carbon dioxide (CO_2) is released.

So how close have we got to ethanol at this point? Remember that ethanol has the chemical equation C_2H_6O (C_2H_5OH). The acetaldehyde molecule has the equation C_2H_5O, so to get ethanol a hydrogen atom must be added, which is easy, since acetaldehyde is what chemists call "tautomeric" with ethanol (that is, its isomers easily change into one another). In fact, all aldehydes are tautomeric with all enols, of which ethanol is one. As a result, to become ethanol all an aldehyde need do is acquire one proton, which comes from the classic proton donor molecule NADPH.

Fortunately for wine drinkers, yeasts have evolved the necessary molecular machines, and it is within yeasts that the various transformations occur when wine or beer is made—or when a baker makes bread. Remember that the byproducts of yeast fermentation are carbon dioxide and ethanol, so when bread is made both a gas (CO_2) and ethanol are given off. The carbon dioxide makes bubbles in the dough, causes the bread to rise, and dissipates during baking. But if ethanol is also given off when bread is baked, why doesn't bread make us drunk? Well, the high temperatures involved in baking bread cause most of the ethanol to evaporate. It has been estimated, however, that freshly baked breads have about 0.04 percent to 1.9 percent alcohol content. The top of this range is about half the alcohol content of weak beer, and a little more than a tenth the alcohol content of wine. If you'd like to get drunk from eating bread, you'll have to eat it right out of the oven. As it cools down, the ethanol evaporates.

There are many other ways in which the reactions involved in fermentation could proceed. Two of these play an important role in winemaking and wine drinking. Yeasts have evolved a specific set of molecular machines and chemical reactions to deal with sugar. Bacteria also turn sugar into alcohol, but by a different process. They too create pyruvate molecules by glycolysis, but they have their own method of dealing with the pyruvate molecules. In the absence of oxygen, or of the enzyme aldehyde decarboxylase (something yeasts have but bacteria do not), the reactive pyruvate will grab an electron from NADPH, to produce NADP. This added

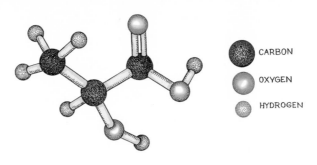

CARBON

OXYGEN

HYDROGEN

Stick-and-ball structure of lactic acid

electron causes the pyruvate to be reduced, and, as represented in the diagram, to change from pyruvate to the small molecule known as lactic acid. Note that the change is in the middle carbon of the pyruvate molecule. What has happened is that the doubly bound oxygen has taken up hydrogen (has become reduced, as chemists say) to form an OH group that sticks out of the middle carbon. This process produces NADP, which can be recycled via glycolysis. So the bacterial cell has found a distinctive, economical way to deal with its electrons.

In a side-by-side comparison of the products that bacteria (left) and yeast, respectively, make from pyruvate, we can see that the two molecules look very different; they taste different as well. Bacterial fermentation is not necessarily bad: humans use it for many food products, including some wines. The USDA requires that two bacteria, *Lactobacillus*

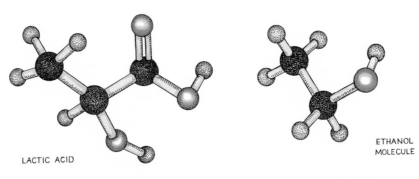

LACTIC ACID

ETHANOL
MOLECULE

Stick-and-ball models of the products of bacterial fermentation (*left*)
and yeast fermentation

bulgaricus and *Streptococcus thermophilus,* be present in yogurt, for example. And other foods with tangy or acidic tastes, such as kimchi and sauer-kraut, also use bacterial fermentation. And, of course, we must not for-get lactic acid itself, an important component of milk and a by-product of several physiological functions in the human body. In some wines, notably many Chardonnays, winemakers might use a secondary bacterial fermen-tation to convert the rather tart-tasting malic acid to lactic acid, which im-parts a more buttery flavor.

It is fortunate that many chemical reactions can be reversed, because even though we may enjoy its impact on our brains in moderate quanti-ties, alcohol is toxic to cells. When we detoxify alcohol in the liver (as will be discussed in Chapter 10), we capitalize on breakdown mechanisms that were probably initially acquired for other metabolic purposes. The simple alcohol molecule is degraded into yet smaller and less toxic molecules through the action of a molecular machine called alcohol dehydrogenase (ADH). Without this particular molecular machine we could not tolerate the toxicity of alcohol, and would not be able to drink wine, beer, or other alcoholic beverages—or perhaps even to eat warm bread.

✦ ✦ ✦

Many textbooks for sommeliers claim that fermentation is as easy as

Sugars + yeast = alcohol + carbon dioxide

If only the world were this simple! This equation omits many components from both the right and the left sides. If the sommelier's job is to know the categories of wine, and what they all taste like—a difficult enough skill to master without the science—then fermentation can be regarded as a black box, holistically. We have simplified our explanation of how fermentation occurs, but we have opted to include some background we consider im-portant to a full appreciation of the life of grapes, yeasts, and other species involved in winemaking.

✦ ✦ ✦

Many of the results of fermentation will have been determined long before the grapes are harvested, when the winemaker chose which grape strain to grow. Winemakers choose grape varieties for their color (deter-mined by molecules), sugar content (also determined by molecules), flavor characteristics (more molecules) and ripening characteristics (ultimately determined by molecules). There are many other considerations as well, most of which involve which molecules will end up in the mixture of crushed grapes ("must") from which the wine is made.

As the grapes are pressed, the sugar molecules come rushing out of the ruptured pulp cells, along with water and other small molecules. Some of the other molecules include the pigments and cellulose-like molecules that reside in the skins of the grapes. The seeds and grape stems may also be caught up in the crush. These release small molecules called tannins into the concoction, along with more cellulose and other molecules that are mostly not active in fermentation. The grape pulp also releases larger molecules such as long-chain proteins and carbohydrates, and some bigger constituents also get into the act—any bacteria or yeasts hanging out on the outside of the grapes will end up in the mixture.

As a result, the "sugars" on the left-hand side of our equation is definitely an understatement. In fact, there are thousands of proteins in the pulp. In one study of grape genes, experimenters asked which of a total of about fifteen thousand genes were making proteins, and found that about 75 percent were. Before crushing, in other words, there were some ten thousand different kinds of proteins swimming around in the grapes, along with the sugars and carbohydrates and other long-chain sugar molecules. Seeds and grape skins were also examined for protein-making genes, with similar results. Enormous quantities of proteins are invariably present in the pulpy, seedy, and stemmy must.

In winemaking, a specific yeast strain is usually introduced into the mixture right after the grapes are crushed. And frequently, yeast and bacteria residing on the outside of the grape skins, or floating in the air, also come along for the ride, and may start to influence what happens in the rich molecular stew. But if there is enough of the added yeast, it is that yeast which will take over. This is, after all, the perfect medium for the growth of yeast cells. So as soon as they are squeezed from the ruptured grape cells, the now unstable proteins begin to degrade, while the yeast cells additionally rummage through the crush and extract whatever they can use. After a while, the only molecules left in the must are small sugars plus larger carbohydrates that are themselves being broken down.

And this is the point at which the right side of the equation starts to make sense. By the time that most of the proteins have degraded into the molecules that the yeast cells can use, their original form has become irrelevant. The sugar- and carbohydrate-loving yeasts busily break down the long-chain sugars into single-ringed sugar molecules, after which the sugar rings are further converted into two small carbon-containing molecules: the ethanol (alcohol) and carbon dioxide in the equation. As long as there is sugar in the must—which depends on how much there was to start with—the yeasts will party on, and produce more ethanol. But once all of the sugars are broken down into alcohol and carbon dioxide, the yeasts begin to starve, stop growing, and die. This may even happen before the sugar runs out, when the accumulated alcohol hits about 15 percent of the must and starts to be toxic to the yeast. The yeasts perish, and no more ethanol is made. This explains why wines are mostly between 9 percent and 15 percent alcohol, and why after fermentation there is a sediment of dead yeast that the winemaker has to remove by filtration, or by racking the young wine from one container to another.

Fermentation by yeasts or bacteria, while crucial, is not the only process by which wine becomes wine. Other molecules, including pigments, tannins, phenolics, and alkaloids, persist even after the sugars are broken down. The pigments give wine its color, the familiar red coming mostly from molecules called anthocyanins, although tannins can also influence a wine's hue. The tannins and pigments are embedded in the skins of the growing grapes, and are harder to extract and get into the must than the sugars.

To produce a red wine, the producer usually leaves the wine on the skins (does not remove them) throughout fermentation, which maximizes extraction. White wines are generally racked straight off the skins, and much of the color of darker-hued whites may in fact come from the oak in which they were aged. But if they are macerated on their skins for a long time, white wines deepen in color, too, as is the case with the "orange" wines made by some bold Italian winemakers, including Paolo Bea in Umbria and Josko Gravner and Stanko Radikon in Friuli. Known by the Italians as "skin-contact whites," these represent a revival of an ancient tradition. The white grape skins are left in contact with the wine for months, during which it becomes opaque and dense with extract. The resulting wines are not for everyone, but there is no contesting that they are among the most complex and interesting around.

The colors of both red and white wines are also impacted when they are matured in oak barrels, which yield a range of molecules to the wines residing within them. At a later stage, both color and flavor also change in the bottle, as molecules continue to interact and break down in the familiar aging process. The aging phenomenon leads to a convergence in color: whites tend to darken with age, while reds become lighter. Without reading the label, one might find it tough to tell what the original color of some ancient wines might have been.

If a rosé is to be made, in most cases wine from red grapes will be left on the skins for a short period before being racked off. Some of the red pigment and flavor molecules contained in the skins can thus make their way into the wine, and the depth of color of the resulting product is roughly proportional to the amount of contact time, typically between one and three days. Occasionally a process known as *saignée* (bleeding) is used, in which pink juice is extracted from red must to increase its concentration. That juice can then be fermented separately to produce a rosé. Other means of obtaining lightly colored wines include blending white and red wines—frowned upon in some places—and co-pigmentation, in which pigments are bound to colorless flavonoid molecules.

Finally, to get a sparkle in the wine—white, rosé, or sometimes even red—it is necessary to trap the carbon dioxide released by fermentation. In the traditional method used in Champagne, and occasionally but widely elsewhere, this is done via a secondary fermentation in the bottle. After pressing, a preliminary fermentation is done in large stainless steel vats, and the resulting still wines are blended as desired. The blend is then bottled, along with yeast and some additional sugar to begin the secondary fermentation. A crown cap is used as a temporary seal. During the second fermentation the bubbles form, and the yeasts die to produce a sediment known as the lees. After the wine has rested on the lees for an extended period, the bottles are gradually moved into an upright position, with the neck at the bottom. The lees collect in the neck, leaving a clear wine above. The bottle necks are then dipped into a very cold brine to flash-freeze the sediment, which is expelled in a frozen lump by gas pressure as soon as the crown cap is removed. At that point the bottle is swiftly topped up with a "dose" that may contain sugar to adjust the wine's sweetness as desired, and the permanent cork is inserted and secured with its familiar wire cage. This is a pretty labor-intensive process, and most sparkling wines today, including the ubiquitous Proseccos, are produced using a "bulk" process whereby the secondary fermentation is achieved in large pressure- and temperature-controlled steel tanks, and the wine is bottled under pressure.

✦ ✦ ✦

Just as a major threshold in the structuring of the universe was crossed when stardust started to form atoms, another major benchmark was passed when molecules began to be replicated, in a process that is basic to life itself. Nature probably tried quite a few ways of doing this before settling on a chemical solution using deoxyribonucleic acid (DNA), the vehicle by which most organic replication is now accomplished. DNA is the molecule of heredity, carrying the genetic blueprint for each one of us between generations; and, at least to biologists, everything about this molecule is beautiful—its shape, its symmetry, its complementarity, and its function.

The long DNA molecule is made up of building blocks, or nucleotides, of four types: guanine, adenine, thymine, and cytosine, known for short as G, A, T, and C. These are arranged in ladderlike twin coiling strands (the double helix), in which each rung consists of a C matched with a G or of an A with a T. It is this constraint that makes DNA the beautiful, symmetrical, and replicable molecule it is: if you have one strand of the double helix, you know what its counterpart will look like. Another aspect about DNA that scientists love is the linear arrangement that results from the way one nucleotide binds to the next. The molecule works by coding for the production of proteins (the building blocks of the cell) in its nucleotide sequence, each coding gene corresponding to a particular protein. And because DNA is linear, proteins that are coded for by DNA are also linear in their primary structure.

Proteins are made up of twenty kinds of amino acid. But there are only four DNA bases, so if each nucleotide coded directly for an amino

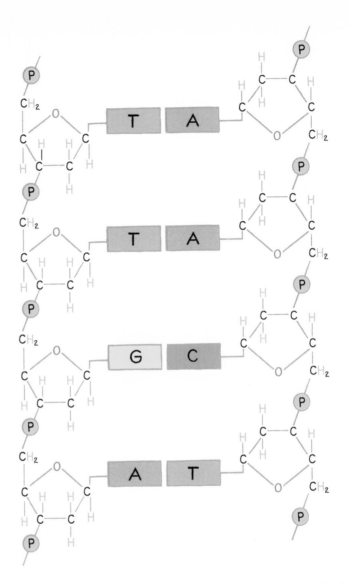

The structure of a double-stranded DNA molecule. The basic building blocks, called nucleotides (G, A, T, C), are shown for four nucleotide pairs. Note that A pairs with T and G pairs with C. The molecule would actually have a helical structure, as determined by Watson and Crick in 1953, but is shown flat here for clarity.

acid, something would be wrong. Even having two nucleotides code for an amino acid would be difficult, because there are only sixteen ways to arrange two nucleotides next to each other. So how about three nucleotides in a row? Three nucleotides yield four possibilities raised to the third power, or sixty-four. Nature therefore provides sixty-four codons, of three nucleotides each. But apparently, when natural selection settled on this triplet code, it didn't mind redundancy, so four DNA codons, CCA, CCG, CCT, and CCC, specify the same amino acid, proline. Up to six codons can correspond to the same amino acid, somehow skipping the number 5.

Just as DNA winds into a double helix and folds itself to produce a higher-order structure, so do proteins. But while almost all DNA likes to wind itself into a double helix, proteins exhibit many and varied ways of folding. And it is the way they fold that gives them the three-dimensional structure that is vital to their function. Through the proteins the DNA codes for developmental processes, and ultimately interacts with the environment to determine the finished appearance of the organism. This is why you look a little like both your parents, and why your children or siblings look a little like you. At a grander level, it is also why cats have much in common with dogs and seals, and chimps, gorillas, and primates in general share many more similarities with one another than they do with other organisms.

Because scientists now understand how DNA and proteins are structured, they have been able to develop techniques that can easily and rapidly decipher both the primary sequences of nucleotides in the genomes of organisms and the amino acid sequences of their proteins. A lot of information thus becomes available, because since the DNA encodes the proteins and enzymes essential for any organism, if the DNA sequence of an organism is known, scientists will already understand a great deal about its characteristics. Beyond this, DNA sequences can also be used to identify either individuals (as in the DNA fingerprinting carried out on crime-scene television) or the species origin of a tissue (in a process known as DNA barcoding). There is an added bonus for evolutionary biologists. DNA has been handed down from parent to offspring (for bacteria, from mother cell to daughter cell) ever since life itself originated, with occasional errors in the replication process (mutations) resulting in the substitution of one kind of nucleotide for another. As a result, this long molecule contains a record of how life evolved. It's worth a moment of digression to see how.

✦ ✦ ✦

How are the different types of grape vines related to one another? Taking his lead from earlier naturalists, Charles Darwin showed in the mid-nineteenth century how the great branching tree of life inevitably results from evolutionary processes. We now know that the structure of this tree is written into the DNA of every organism on earth, though some are yet to be deciphered.

All life on earth has diversified from a single common ancestor, in a sequence of successive branching events. Every group of organisms shares a common ancestor, and that ancestor in turn shared a more remote common ancestor with related groups. The tree image perfectly illustrates this process, allowing those common ancestors to be reconstructed. This step is crucial in understanding the identities of the players in the making of wine (as we'll discuss further in the next chapter).

Let's look at a simple example, in which we have a grapevine, a rose, a corn plant, a ginkgo, and moss. These photosynthesizing organisms are all plants, and their relationships are uncontroversial. Among them, grapes and roses are the most closely related, as we see in their unique shared form of embryonic development. Next comes the corn, equally related to roses and grapes in a group united by having flowers. Equally related to grapes, roses, and corn is the ginkgo, in a larger group whose members all produce seeds. And this, of course, leaves the mosses as the outlier, related equally to all the others.

The illustration shows that we can represent these relationships using a branching diagram. When two things are each other's closest relative, they are drawn as two branches connected at a fork. There are several forks in the tree, which are numbered here and each of which represents the common ancestor of the organisms above it. Thus the fork numbered 2 in the figure can be thought of as the common ancestor of corn, rose, and grape. Once we have determined that there was a common ancestor such as the one for roses, grapes, and corn, we can ask some important questions. How ancient might that ancestor be? Are there any known fossils

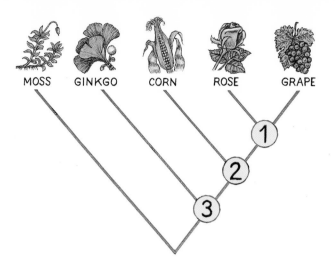

A phylogenetic tree of plant relationships, illustrating the relationship of roses and grapes

that might coincide with it? If we can answer these questions, we will know how old the group that contains rose, corn, and grape is. We can also ask what the common ancestor looked like, and how it functioned.

The evolutionary tree has a respectably long history. One of the most iconic branching trees in all science is Darwin's "I think" tree, scrawled in one of his notebooks when he was twenty-eight years old and fresh off his round-the-world voyage on HMS *Beagle*. It is now also widely found tattooed on the persons of evolutionary biologists.

Still, techniques of tree-building have come a long way since Darwin's time, and several methods have been developed for constructing trees based on DNA sequences. The simplest approach is the example we just gave of the grape, rose, corn, ginkgo, and moss, in which simple similarities were used to group taxa (units) together. Other approaches are needed, though, because simply having a similar appearance does not prove that organisms are closely related. Examples abound in nature. One of the neatest occurs in plants, where the euphorbs, a group of plants living in Asia and Africa, have converged on the New World cacti. They are very similar to each other in appearance, yet in terms of their evolutionary histories they are only distantly related.

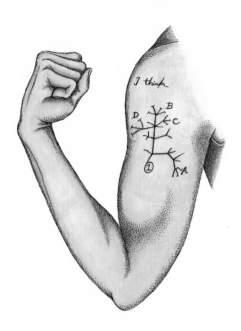

Charles Darwin's "I think" idealized evolutionary tree of 1837

In building our trees, we thus have to abandon overall similarity and look specifically for features that were inherited from a common ancestor, rather than acquired independently. Because the long DNA nucleotide chains inevitably vary among species in the bases that constitute their links (as the result of mutations in a succession of common ancestors), DNA is an ideal tool for this job. If we have grapes, roses, and corn, and we want to know how to arrange them based on DNA sequence data, we can sequence a gene for all three species. But a frame of reference is needed. Imagine that grapes and roses both have an A (adenine) in the last position of the gene, while corn has a T (thymine) in this position. We might immediately conclude that the grape and rose were sisters; but in fact, without some larger frame of reference all we really know is that there was either a change from an A to a T or a change from a T to an A somewhere in the evolution of these three plants. To help us resolve the direction of the change, we might look at the ginkgo. To find the best tree for these species we start by seeing how the base pair change maps onto all the possible ways there are to arrange grapes, roses, and corn. In this case it's pretty

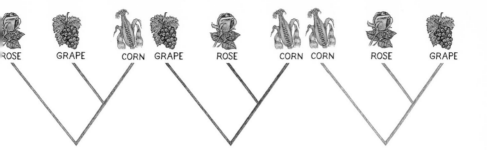

ROSE GRAPE CORN GRAPE ROSE CORN CORN ROSE GRAPE

Three possible phylogenetic trees for corn, roses, and grapes

simple, because there are only three possible configurations. We can put grapes with roses, grapes with corn, or corn with roses. These three trees are shown in the figure.

Imagine that the ginkgo is sequenced, and it contains an A in that last position. In that case, the best explanation for the T in corn is that it is a novel mutation in the lineage leading to corn, which wouldn't help us decide on the arrangement of our target species. But if the ginkgo turns out to have a T, we can use the principle of parsimony to judge which of the three possible trees is best—that is, the one that most efficiently explains the data. Looking at the tree that has grapes and corn as closest relatives, we find that the DNA sequence has to change in two places. This is also the case for the diagram that places the roses with the corn plant. But the tree in which grapes and roses go together requires only a single change to explain the sequences observed—so it is the "best" tree for the small data set we have.

Of course, to do a study like this properly we would have to look at thousands, if not millions, of DNA sequence changes, as Ernest Lee and his colleagues did at New York University in 2011. These researchers examined over two thousand genes, and found that more than five hundred DNA sequence positions supported the roses-with-grapes tree, while only around thirty supported either of the other two possible hypotheses.

4

Grapes and Grapevines

An Issue of Identity

O ur quest to sample a wine made from the most ancient grape
variety we could find had led us to an inexpensive sparkler from
southern France whose predecessors Pliny the Elder had praised two
thousand years ago. The Clairette that now stood on the table in front
of us reposed in a chilled green bottle capped with faux gold foil, and
the cork came out with a satisfying exhalation. Large, lazy bubbles
surprisingly resolved into a light, creamy fizz on the palate, followed
by delicate sweetish tones of honey and melon. A perfect aperitif for
a warm summer's evening. The Romans had evidently known what
they were doing.

Take a glass of wine in your hand, and hold it up to the light. Then
swirl, sniff, slurp, and swallow. At every stage of this ritual your senses
will be entertained: sight, smell, taste, touch—even hearing, if you slurp
vigorously enough. For a liquid with such a simple appearance this multi-
sensory appeal might seem remarkable. But wine is a complex concoction, and making wine requires combining many species of organisms into
an intricate microbial ecosystem within which numerous delicate interactions take place. For such a subtle and nuanced product one might expect no less, and it is hardly surprising that an understanding of how wine
is made takes more than a knowledge of how yeast transforms the sugars
in grapes into alcohol via a sequence of chemical reactions. This may be
the fundamental process in winemaking, but a lot more is going on. What
happens as grape juice is transformed into wine is rooted in the life histories of both grapes and yeast.

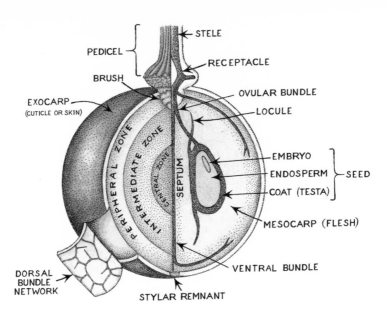

Cross-section of a grape

✦ ✦ ✦

Let's start with the grape. A basic wine grape consists of nothing more than embryos—seeds—surrounded by a thick fleshy casing encapsulated within a thin, tough skin. Some grapes have been bred to be seedless, through changes in the genes that code for making seeds. This is the case with most table grapes because seeds are not pleasant to munch on and have a bitter taste. Although attempts have been made to make wine with seedless grapes, no reputable seedless wine grapes currently exist.

Within the waterproof skin, the flesh encasing the seeds contains minuscule interconnecting tunnels that circulate nutrients, hormones, and water. There are usually four seeds in a wild grape, but this number can vary. Each seed is made up of a soft, fleshy embryo surrounded by a membrane called the endosperm, itself surrounded by the hard seed coat, as shown in the illustration.

At the end of each grape there is a stem, or pedicel, that connects it to the vine. The pedicel is a kind of valve through which nutrients and water pass to provide the grape with nourishment. Like our own bodies, plants need to transport nutrients to their various constituent parts. The major

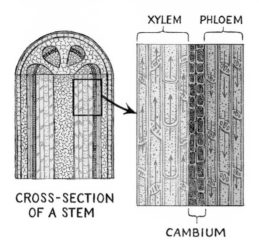

XYLEM   PHLOEM

CROSS-SECTION
OF A STEM

CAMBIUM

Position of the xylem and phloem in the stem of a typical vascular plant
(the cambium is part of the secondary vascular system)

nutrient carrier in our bodies is the blood that circulates through our vascular system of veins, arteries, and capillaries. In plants, two vascular systems pass through the pedicel, both made up of grouped cells that form the tiny tubelike structures called xylems and phloems. One tube typically lies inside of the other, forming a tube-within-a-tube system.

The xylem and phloem act as sieves, controlling the flow of molecules such as proteins, sugars, and hormones into and out of the grape. The xylem is the inner tube and is responsible for transport of water, growth hormones, minerals, and any nutrients that come from the root system that feeds the remainder of the vine. It plays an important role during early development of the grape, but it loses its importance as the grape continues ripening. The later stage of grape development is known to grape growers as *véraison* (onset of ripening), and its beginning is usually announced by the shutting down of the xylem vascular system.

The phloem is responsible for the transport of sucrose and products of photosynthesis in the leafy area of the vine. Before véraison, phloem activity is muted, but once véraison commences, the phloem comes into its own. Together, the xylem and phloem regulate the size and volume of the grapes, and hence their sugar and water content. Controlling the quantities of these critical ingredients in each grape is essential in fermentation

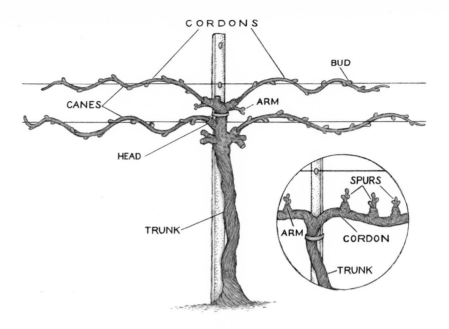

Vinous parts of a grapevine

and central to winemaking: the sugar and water are what is turned into alcohol. But since almost all the components of the grape—pulp, seed, and skin—have a role to play in winemaking, getting all these elements in balance is also critical to the look, taste, and feel of a wine.

The structure of the vine itself is also important for understanding both how the grape gets its nutrients (which are key to its quality) and how the vine can reproduce asexually. The vine has deep roots that anchor each plant in the ground and serve as conduits for the nutrients absorbed into the plant from the soil. The roots and lower part of the trunk are called the rootstock. Above this, the trunk extends up to the head, which in turn gives rise to numerous branchlike extensions called cordons. On each cordon are budlike bodies that develop into canes. Dotting the canes at regular intervals are smaller buds that swell and develop into leaves and grape bunches. The intervals are called internodes, and the numbers of buds and internodes are key to the annual pruning of the vine. After the dormant period (usually during the winter), the buds start to swell. A green shoot tip emerges after the swelling, then the leaves start to de-

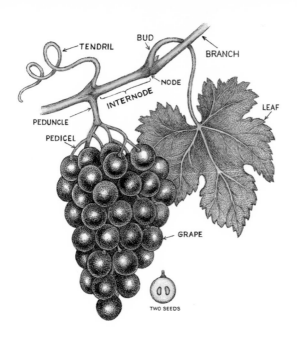

Anatomy of the terminal end of a grapevine

velop. They unfold, and grape flowers begin to grow and separate from them. The flowers bloom and are fertilized, the fruit sets and begins to develop. When mature, the grapes can have one of several fates: harvesting by humans, consumption by animals, or rotting. The leaves then fall from the cane, bringing the cycle back to where it started.

Breeders have energetically modified virtually every feature of the vine through stringent selection of plants, and spent thousands of years training grapes to produce characteristics that meet winemakers' precise specifications. The process is in principle much like the breeding of animals, which also has a long history; but domesticating grapes has turned out to be more like herding cats than breeding cattle. All plants are difficult to "train," and in the end the best solution proved to be waiting for spontaneous changes in the genome that impacted how certain desired traits developed.

As discussed in Chapter 3, DNA is the long, spiraling double chain of molecules that carries the information passed on from generation to generation and decoded to direct the development of each individual. The genome of animals, yeasts, and plants is the totality of the DNA in the nucleus of a single cell, and it is bundled into the chromosomes that contain the genes. In the human genome there are 23 pairs of chromosomes, 20,000 genes or so, and 3 billion nucleic acid pairs (the guanines, adenines, thymines, and cytosines). Two copies of the genome (one from the mother and one from the father) exist in almost every cell in the human body. In grapevines there are 19 pairs of chromosomes and 26,000 genes, although the genome is only 500 million nucleic acid pairs long. Yeasts differ in having only 16 pairs of chromosomes, about 6,000 genes, and a mere 12 million nucleic acid pairs.

Between 1856 and 1863, Gregor Mendel, a monk living in what is now the Czech Republic, conducted experiments on the flowering peas in his monastery garden that enabled him to discover the basic rules of genetic inheritance. After nearly a century and half of genetic research on organisms such as flies, worms, sea slugs, mice, and thale cress weeds, scientists have added to his findings about how traits arise and are inherited. Unlike the features of the peas raised by Mendel, which were simple in their inheritance patterns, most traits that we see in grapes (and, for that matter, in most organisms) are what geneticists call complex, controlled by many genes, rather than just one. And with the addition of DNA sequencing techniques to the geneticist's arsenal, the changes in the DNA of an organism can now be determined for both the simple and the complex features we observe in plants, fungi, animals, and microbes.

Armed with this knowledge, modern plant and animal breeders since the 1970s have begun to deploy the techniques of genetic engineering to develop new, beneficial traits. But even without these techniques, breeders were able to target many traits that are easily observed and clearly heritable. Plant breeding has a rich history, and has resulted in a staggering number of grape varieties (or strains) that have been produced through the many different ways plant reproduction and growth can be manipulated. The simplest and most common approach—and indeed the method that nature used before humans came along—employs sexual reproduction.

Wild grapevines have what is called a dioecious sex life: individual plants are strictly male or strictly female, and offspring may show capricious combinations of parental traits. Because there is a male contribution and a female contribution to offspring plants, the traits of both need to be tracked in order to predict probable traits in the offspring. This is the job of the geneticist or breeder. Early on, geneticists realized that there were two major kinds of traits—those that were clearly expressed in every generation and those that skipped a generation. They began to call the regularly expressed traits dominant and the traits that skipped a generation recessive. Cultivated grapes, however, have been bred to be more manipulable than wild grapes. Their flowers possess both male and female parts: they can mate with themselves and still produce fertile offspring. The term for this kind of sex is *monoecious,* and over breedings the process tends to homogenize the genetic makeup of offspring.

Yet another way to make grapevines reproduce and give a desirable product is to hybridize two kinds of grapevine, each with a different desired trait. Plants can do this because the number of chromosomes present during the union of the ova and pollen is immaterial to the breeding. In this respect plants differ from animals, which must have matching chromosomes when the sperm and egg unite. Specifically, animal zygotes (the union of an egg and sperm) require there to be the same number of chromosomes in each. Hybridization between divergent plant species is thus much easier than in animals. If the hybridization goes as planned by the viticulturist, the new hybrid offspring grapevine will exhibit the desired divergent traits of its parents, and subsequently can be widely propagated.

At the same time, there are several ways of strictly maintaining the characteristics desired of a particular kind of grape. The first is to clone the grapevine, bypassing the reproductive organs. In principle, this is as simple as cutting canes from a vine and planting them in the ground, or air-layering the grapevine, an ancient system based on inducing roots to form on a vine stem without detaching it from the plant. But in practice there are complications. There is something weird about plant cells. In early development, animals have stem cells that start out not knowing what they want to be but eventually differentiate into many different kinds of cells—neurons, skin cells, and so forth. The stem cells' ability to become any kind of cell makes them pluripotent. But once an animal stem cell has decided on its role in life it becomes fixed and loses its pluripotency. In contrast, even when differentiated into specialized cell types, plant cells retain the potential to develop into other kinds of tissue. It is a simple procedure to use one plant part to regenerate the whole.

Usually when cloning a plant the viticulturist takes cuttings (for grapevines, canes are preferred) and soaks them in plant hormones that will induce roots to grow. These treatments activate the root-generating genes. The rooted cutting is then planted, and since the cloning does not involve sex, the resulting vines are genetic replicas of the parent. In this way, grape growers can ensure the exact reproduction of any vine that will cooperate with the cloning process, and this in turn ensures great consistency among the vines—and therefore the grapes—in a vineyard. If all the vines are clones of one another and their environment is similar, they ought to produce an effectively identical product. If the viticulturist opts for layering, a cane from one plant (called the mother) is stretched down to the ground—rather resembling an umbilical cord—and buried, leaving only the tip of the cane and its buds aboveground. This extended cane usually takes a year or two to reach a trunk diameter bigger than the "umbilical cord" of the mother cane, but when it does the mother cane is snipped and the daughter plant allowed to grow on its own.

Another method growers use to ensure genetic uniformity among their grapevines is rootstock breeding. This technique involves an already well-established plant with a healthy root system—either a whole plant or, more commonly, a stump. A cutting—the scion—is made from a vine with desirable characteristics and grafted onto the stump. The scion usually fuses nicely with the rootstock, and over time fusion becomes complete so that the two parts behave as a single plant. But the rootstock has one set of genes—which are usually chosen to enhance root growth or resistance to pathogens—and the scion has another set, which usually controls the characteristics of the fruit. This particular property of grapevines—that they can be grafted—has also been exploited for purposes that go far beyond breeding: grafting was, for example, critical to rescuing the wine industry in the nineteenth century from the depredations of the phylloxera insect, as we discuss in Chapter 7.

So which traits are the most desirable for making a good wine? The sugar in a grape is important, because it is the major fuel the yeast uses for

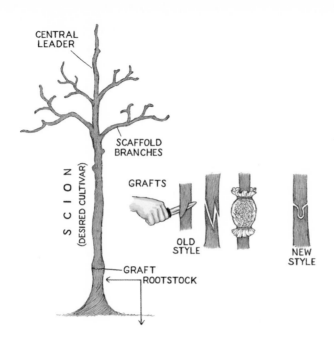

CENTRAL
LEADER

SCAFFOLD
BRANCHES

SCION
(DESIRED CULTIVAR)

GRAFTS

OLD
STYLE

NEW
STYLE

GRAFT
ROOTSTOCK

Diagram of a scion, showing the position of the rootstock relative to the graft

making alcohol, so we might focus our breeding on the amount of sugar in the pulp. And because one of the major hallmarks of good wine is consistency, we might also want to make sure the sugar content is relatively constant in the pulp within each strain. We would additionally want our grapes to breed true from one generation to the next, though perhaps to develop larger leaves. This trait would increase photorespiration, which would in turn produce more sugar for the grapes. And let's not forget bunch size. A grapevine that produces two or three times as many grapes per bunch as the average would give more yield per vine, although excessive productivity lowers the quality of each grape, whereas severe pruning generally ensures a better wine. In the case of a complex product such as wine, the list of potentially desirable qualities is long. All these desirable traits, and many more, have been bred into cultivated grape strains, resulting in a huge number of varieties.

Let's return for a moment to grape seeds, which are important not only for understanding the propagation of grapevines but also for appreciating how different traits in grapes have been selected. Before grapes were cultivated and bred by people, the normal number of seeds in a grape was four. But by manipulating the grapes' genes, plant breeders were able to grow grapes without seeds. (Seedless grapes actually do have seeds at an early point in their development, but the hard outer seed coat fails to develop as a result of genetic mutation.) How did grapes become seedless? Genes consist of DNA. When parents produce eggs and sperm, the amount of DNA within them that is passed on to their offspring is reduced by half, though it is usually passed along intact. But the sequences in the egg and sperm are not always identical to those of the parents because sometimes, though very rarely, slight mistakes are made in copying the DNA. These mutations are not always detrimental to the organism. Indeed, they may have no impact on the organism at all, and they are ubiquitous in the living world.

How do mutations occur? Imagine that you have a set of instructions for a specific task you need to perform, such as "Start making seed coats." Changing some of the words in the instructions—for example, "start to make seed coat," or even "start making seed coat"—would often not change their functional meaning. But if the word *start* is changed to *stop*, then the message becomes entirely different. Observing mutations in living organisms is an art, and some researchers have made their reputations on their ability to recognize and isolate mutants in a broad range of organisms. Ironically, since nobody has yet been able to produce a seedless grape variety that makes a decent wine, some of the most spectacular successes in understanding plant genetics have come from locating and characterizing the gene responsible for seedlessness in fruits; but if similar techniques can be used to manipulate grapevine genes for sugar content, grape color, or other important traits, useful advances may well be expected in how we make wine—and consequently in how we drink it. Such work is still on the horizon as far as vines are concerned, but odds are that new discoveries will soon add to traditional selection procedures and help produce an even wider variety of kinds of grapes. And now, because the place of the grapevine in the plant world is essential to understanding its characteristics (and why it is able to provide humans with such gratification), we turn to how grapes and their wild cousins are related to each other, and to the rest of the plants.

✦ ✦ ✦

Life probably began in the oceans. But after the first plants, animals, and fungi had established themselves on land, a little under half a billion years ago in the Cambrian Period, a process known as adaptive divergence began. Adaptive divergence occurs when a burst of new forms appears, and opportunistic organisms fill a wide range of newly available ecological niches. It seems that adaptive divergence on a variety of scales has been one of the more important processes in generating the luxuriant diversity of life on our planet.

One of the main accommodations plants made to terrestrial life involved the restructuring of their body plans. Specifically, the early land plants developed a vascular system that allowed for the internal transport of water, leaving plants without vascular systems to the more primitive aquatic life, or to specialized terrestrial niches. Such nonvascular plants survive today in the form of green algae and the bryophyte group that includes the mosses, liverworts, and hornworts, all solitary reproducers.

Although a botanist who studies mosses and liverworts would vigorously disagree, to us it appears that vascularization is where the action really began in plant diversity and adaptive divergence. And by looking at strange species that have changed barely at all over several hundred million years of evolution, researchers can hypothesize about how vascular plants diversified and gave rise to such remarkable descendants as grapevines. Popularly called living fossils (although our colleague Richard Fortey prefers to describe them as survivors), these hardy species include odd vascular forms like clubmosses, horsetails, ferns, ginkgoes, and cycads.

Clubmosses are perhaps the most primitive vascular plants, followed by the horsetails. These plants have almost exactly the same anatomical structure they had half a billion years ago. Ferns appeared in the fossil record about 350 million years ago, but most of their current diversity can be dated back to fossils that existed about 145 million years ago. And while clubmosses, horsetails, and ferns do not form a natural group, they do share a trait that is unique: they do not generate seeds but instead reproduce by making spores. All other vascular plants produce seeds for reproduction, and today's floras contain two great lineages of seed plants that diverged from each other around 300 million years ago. What separates these two large groups is whether, like vines, they flower (whether they are angiosperms) or not (are gymnosperms).

Most of the primitive vascular plants that have been regarded as living fossils are gymnosperms. Few are flowering angiosperms. In an 1879 letter to the botanist Joseph Dalton Hooker, Charles Darwin, the father of evolutionary thought, called this distinction the "Abominable Mystery." The abominable part of the mystery was that the angiosperms appeared abruptly in the fossil record, just before the orgy of adaptive divergence that produced the bewildering array of flowering plants we see today. This burst of adaptation violated Darwin's expectation that evolution would prove to have been a process of slow, incremental change and diversification, and it flummoxed him. Even today, when we understand that events of many different kinds can contribute to the evolutionary story, it remains remarkable that so many flowering plants contrived to evolve in so short a time (evolutionarily speaking). The first angiosperm fossils are dated to about 135 million years ago, although there are hints from fossil pollen grains that angiosperms might have existed as much as 250 million years ago, vastly upping the divergence time. But the mystery resides not just in the timing. It also lies in how so many different forms could be generated genetically and developmentally, and in why some angiosperm groups diversified wildly, while others failed to do so.

✦ ✦ ✦

So where do grapevines fit into the plant evolutionary tree? Under the conventions of classification, every living species belongs to a group that in turn belongs to a yet larger group. Most wine grapes belong to the species *Vitis vinifera*, which is grouped with other species into the genus *Vitis*, which in turn is classified along with other genera in the family Vitaceae. This family belongs to the order Vitales . . . and on up, through the angiosperms and the Plantae to the domain Eukaryota (which contains everything with a nucleus in its cells, including us).

Vitales itself has proven difficult to classify among the angiosperms in general. Anatomical studies suggest that it groups with the rosids, an extensive assemblage of plants that embraces more than a quarter of all angiosperm species. But the results of genome comparisons have confused the classification, and our best conclusion is that the order Vitales descends from a common ancestor of all the rosids. This common ancestor evidently had huge evolutionary potential, since it eventually gave rise on one hand to roses, among the most beautiful plants, and on the other to the grapes, which give us wine.

Vitales contains only the single family Vitaceae, which is considered distinctive enough to have an order of its own. The family itself is divided into two major groups: the herby and treelike Leeoideae, and the clinging, climbing vines of the Vitoideae. As its name suggests, the grape-bearing genus Vitis belongs (with thirteen other genera) to the latter. Although those thirteen genera are also vinelike, none as far as we know produces fruit suitable for fermented beverages. The genus Vitis itself contains around sixty species. Wild species within the genus Vitis are found for the most part in the northern hemisphere, with varieties occurring in Asia, North America, and Europe.

Not all grapes are equal. Or at least, not all grapes are the same in their potential for making wine. To the taxonomists who attempt to give order to the vinous species, a grape (*Vitis vinifera*) . . . is a grape . . . is . . . not necessarily a grape. Taxonomists follow the rules set by Carl von Linné (Linnaeus) about 250 years ago. At first glance these rules seem simple. Linnaeus gave each species a binominal designation (a combination of two names). The first name denotes the genus to which the species belongs, and the second denotes the species itself. For example, in this system we humans belong to the genus *Homo,* and to the species *Homo sapiens.* So far so good. But beyond these basic rules things can get a little complicated, since taxonomy is traditionally based on expert judgment, and names based on subjective expertise may often change.

Species are, in principle, the largest freely interbreeding populations of organisms, defined by reproductive exclusivity. Practical problems may arise, though, if reproductive isolation is incomplete, or if the taxonomist has to judge reproductive exclusivity purely from the physical appearance of the subjects. And usually the reproducer has a history that complicates matters. A taxonomist might decide, for instance, that something traditionally regarded as a separate species should actually be included with

Vitis vinifera Subspecies

Subspecies	Cultivars	Rootstocks	Hybrids
aestivalis	0	0	++
amurensis	+	0	++
berlandieri	+	+++	0
candicans	0	+	0
caribaea	0	0	+
champinii	+	+	0
cinerea	0	+	++
cordifolia	0	+	+
labrusca	+++	++	+++
longii	+	++	0
riparia	++	+++	+++
rupestris	++	+++	+++
simpsonii	0	+	0
vinifera	+++++	+	+++

its closest relative(s) in the same species. To distinguish the populations from one another, each may then be classified as a subspecies, and will receive a trinomen, a third (subspecies) name at the end of the two-part species name.

Such was the fate of the species *Vitis vinifera,* within which infraspecific names have proliferated over the past century or so, and all the subspecies are interfertile. We can only apologize for the resultant complications, but understanding their origin is essential if one wishes to navigate the maze of names in the world of grapevines and wine. The table lists the most important subspecies of *Vitis vinifera* and their importance to the viticultural industry in various roles. When a zero appears in the table, it indicates that the subspecies is not used for the function listed, while plus marks indicate its degree of utility for the specific function. Thus what used to be the species *Vitis berlandieri* is now known as the subspecies *V.v. berlandieri.* It is used only rarely as a new cultivar (domesticated form), and never in hybridization, but it is a common rootstock.

Vitis vinifera has the distinction of having been named by Linnaeus himself, and if we were concerned only with binominal nomenclature, this name would be all we needed. But in the late nineteenth century the German botanist Carl Ernst Otto Kuntze decided that the taxonomy of *Vitis vinifera* needed some modification. In 1891 he published a tome on plant taxonomy, *Revisio generum plantarum,* in which he renamed thousands of species of plants, including grapevines. This work seems to have rather annoyed the botanical community, and his renamings, like the *Revisio* itself, were largely ignored or rejected by his contemporaries. Nonetheless, his infraspecific names for certain varieties of grape have made something of a comeback, and many of the names in the table are still widely considered valid.

Nor was Kuntze's work the limit of the complications added after Linnaeus departed the scene. Nowadays, taxonomists recognize only about 60 separate species in the genus *Vitis,* but in the more than 250 years over which grapes have been named, some 500 different species names have been given to them. This over-naming usually happens when someone mistakenly gives a new name to a species that already has a name. When a plant is found to have more than one name, taxonomists use the rule of priority to decide on the appropriate name, assigning the first name used to the species and banishing all other names to the scrapheap of biological history. In part, the confusion among grape names is due to the wide variety of variously related strains, but it's also true that scientists tend to want to name the things they study.

To add to the messiness of the taxonomic situation, more than one potential name exists for the wine grapes of greatest interest. The German botanist Henry K. Beger, for example, gave the name *Vitis vinifera sativa* to a particular kind of vine that was later determined to be the same as Linnaeus's *Vitis vinifera* L. ("L" denotes it as a species described by Linnaeus, who by definition has priority). This should mean that Beger's grape, if it is truly the same subspecies as Linnaeus's vine, ought to be known by the trinomen *Vitis vinifera vinifera.* But this name rarely appears in a database search, and when researchers use it they almost certainly mean the vine Beger called *Vitis vinifera sativa* (that is, *Vitis vinifera* L.). To make the situation more complicated, *Vitis vinifera sylvestris* is also used to describe some wild strains of grape that are closely related to cultivated grapes. According to *Plant List,* the web authority on plant nomenclature, this subspecies name is also synonymous with the full species *Vitis vinifera* L. (Adding to the complication, *sylvestris* is often spelled *silvestris.*)

If the various name changes strike you as confusing, we admit we're equally at sea. And there is yet a further complication. Within any species or subspecies in which a lot of variation occurs, scientists might create yet another category. In the case of grapes, this additional category reflects that vines have been domesticated and bred into a host of what are often called varieties. But when we use this term to differentiate among the varied strains of grape that have been cultivated throughout history, we are not referring to the same thing as when we describe "varieties" among wild organisms. So another way to refer to the cultivated forms is to call them accessions, alluding to the way they are catalogued in the canonical American and French reference collections. Technically speaking, we should refer to the domesticated forms (Chardonnay, Syrah, and so forth, and even strains within them) as cultivars, although in the wine literature this term is used fairly interchangeably with "variety" or "varietal."

✦ ✦ ✦

The United States Department of Agriculture tends to work quietly behind the scenes, monitoring the quality of the foods we eat and keeping a tight watch on the health of agricultural crops and animals. But it also maintains one of the world's most important repositories of grapevine varieties: the Cold-Hardy Grape Collection, over 1,800 strains of grape clones, at the Clonal Repository Farm in Geneva, New York. Its size makes the Clonal Repository Farm one of the most important "farms" in the United States. But it pales in comparison with the French Institute for Agricultural Research at Vassal, near Montpellier (which at this writing is, alarmingly, threatened with eviction from its premises by—of all things— a large wine producer). Vassal holds the single most impressive grape collection on earth, with over 4,370 grapevine varieties. Most (almost 4,000) of these are cultured variants of plants classified as *Vitis vinifera*; the rest are hybrids, wild grape strains, and rootstocks.

It has been estimated that there are between six and ten thousand strains of Eurasian grape in the world, and these collections store and maintain more than half of them. The availability to researchers of this wide range of strains, together with recent huge advances in DNA sequencing technology, has led to a revolution over the past five years in our understanding of grape relationships. Two questions about grapes have been approached using the new genetic techniques: What are the immediate ancestors or relatives of the cultivated strains of grapes, and Which strains come from where?

Each of these questions has its own complexities. For instance, most of the cultivars in the French and American collections do appear to trace back ultimately to a relatively recent progenitor strain. But this progenitor might not have been the actual grapevine that produced the earliest wine. In addition, ongoing interbreeding among vine lineages can complicate the tracing of genealogical relationships. When this happens, the relationships in a family tree start to look untidily bushy. So it is fortunate for tracing genealogy that, once a domestic strain is established, it is usually propagated by cloning or grafting (thereby eliminating some of the extreme bushiness in genealogies caused by cross-mating) and that the breeders who implemented many hybridizations of grape cultivars in the past century created records of their work.

At least in principle, genetic studies using a variety of genes and approaches can pinpoint both the identity of the wild progenitor of grapes and how the myriad grape strains are related to one another. With the introduction of faster genome-sequencing techniques at the beginning of the twenty-first century, a sort of international competition broke out to unravel the ancestry of *Vitis*. Wine has a strong European connection, so several laboratories in European Union countries have been racing to determine the closest wild relative to *Vitis vinifera* L. (In human terms, this would be equivalent to asking one of the genealogy companies you find on the Internet to trace your ancestry back to ancient Mesopotamia.)

To start, the labs would use written documentation and old-fashioned detective work. When the trail went cold, and the written records reached a dead end, they would turn to the methods of modern genetics. And since the written record of the crosses that produced most grape cultivars fades out a mere hundred or so years ago, enter the DNA detectives. Using DNA sequencing methods, wine researchers have approached the genealogical question as a part of a larger overall project to understand the genealogy of flowering plants.

In overlapping analyses, Dorothee Tröndle and her colleagues in Germany and Giovanni Zecca and his colleagues in Italy each examined nearly

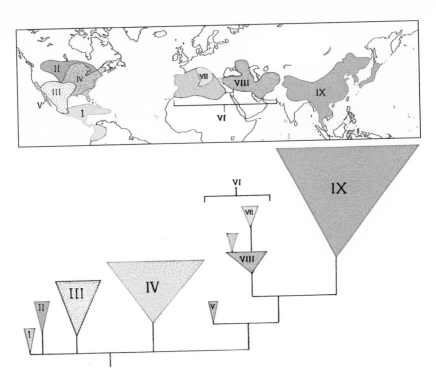

Phylogeny of wild grape strains. Each triangle represents a geographic region of the world as indicated. Each large triangle is made up of several grape strains that are related to one another through common ancestry. The triangle without a label refers to vines from North Africa. Redrawn and modified from Zecca et al., "The Timing and the Mode of Evolution of Wild Grapes (Vitis)."

half the sixty species in the genus Vitis, plus several closely related genera. At a closer level of detail, teams led by José Miguel Martínez Zapater in Spain, Valérie Laucou in France, and Sean Myles in the United States all examined relationships among numerous wild and domesticated grape strains by examining Vitis vinifera L. cultivars, and Vitis vinifera sylvestris accessions.

The Zecca and Tröndle studies came to much the same conclusion. European Vitis vinifera is most closely related to the Asian Vitis species as a group, with the North American Vitis species being odd man out. (Curiously, as there are few South American Vitis, most of the Vitaceae that naturally occur there belong to the genus Cissus.) And when genes are used to look at relationships among the grapes grouped as Vitis, it becomes clear that all come from a single common ancestor, validating the traditional classification. So one assertion we can make right away is that Asian and European grapes are closely related. The Zecca study made two further observations. First, it detected a close relationship between a small subset of Asian *Vitis* species and all North American *Vitis*, suggesting that the North American *Vitis* arose as a result of divergence from this small subset. Asia is thus evidently important to an understanding of how *Vitis* diverged. The study also turned up a considerable degree of intermixing of species within the three continental areas, leading the researchers to conclude that grape evolution has been and continues to be a fluid process.

Reproductive isolation between species involves the end of meaningful genetic contact among them. Any mutation occurring in one lineage after divergence should not show up in others, and thus a mutation can be used as a marker of common ancestry for members of that lineage. The same mutational process is occurring in all the other lineages under consideration, each of which will accrue sets of identifying, unique mutations. But as we have seen, species and populations may have differing degrees of reproductive isolation, and this is apparently the case among the many species of *Vitis*. In other words, reproductive isolation was not complete as new lineages of *Vitis* diverged. Isolation did, however, lead to lineages different enough anatomically to be classified as separate species.

The key question is which of the non-*vinifera* species is most closely related to the large number of *Vitis* cultivars or variants produced since the grapevine was first domesticated. The Zapater group concluded that, among all wild grapevines, *Vitis vinifera sylvestris* was the closest relative of all the wine cultivars. Because *Vitis vinifera sylvestris* is found all across Europe, the Spanish researchers tried to pin down which European wild accession was the original cultivar for all modern wine grapes. As we have seen, the archaeological evidence suggests that winemaking probably began somewhere in the Caucasus, or possibly in nearby Anatolia; and the genetic analysis confirmed a Caucasian–Near Eastern origin for the grape involved. But it also pointed to a second origin, in western Europe. As proud Spaniards, the Zapater group hence concluded that more than 70 percent of the cultivars grown in the Iberian Peninsula are best explained as descendants of western European wild *Vitis vinifera sylvestris*.

But before we adopt this theory, we should note that a French research group, headed by Patrice This, has questioned the Spanish claim. This and his colleagues explored whether these putative Spanish ancestors of all wine grapes were "real *sylvestris* individuals that have never undergone cultivation, or 'escaped' individuals from vineyards or hybrids between wild and cultivated forms." They proposed that, using plant genotyping tests (rather like the paternity tests used as evidence in courts), researchers could answer this question. Accordingly Valérie Laucou and her colleagues, also in France, changed the focus of research from identifying the progenitor of wine to unraveling the relationships among different wine grape varieties. They used an approach called microsatellite analysis to evaluate the 4,370 grape strains found in the Vassal collection, of which 2,300 or so were cultivars. Their data set also included wild strains, hybrids, and rootstock specimens. And they used the same DNA fingerprinting technique that is used on television crime programs like *CSI* and *Dexter* to determine whether an individual was present at a crime scene.

◆ ◆ ◆

DNA fingerprinting is a simple approach that is a little like counting the stripes on a zebra or the whorls in a human fingerprint. Within the genomes of organisms, there are often nonharmful changes in which a particular small sequence (2 to 6 nucleotides long) is repeated many times. So, for instance, in one individual a specific gene region might have a sequence of ATATATATATATATATATAT (AT repeated eleven times) inserted into it. In another individual of the same population, the corresponding region might have ATATATATATATATATATATATATATATAT (AT repeated fifteen times). These repeats can be isolated, using DNA sequencing machines that characterize them as bands or stripes, based on their size or number of repeats. The positions of the stripes correspond to the number of repeats in the DNA of the gene. It has been estimated there are thousands of these small repeated "microsatellite" regions distributed throughout most eukaryotic genomes, and they change rapidly from generation to generation, so a lot of variation occurs even among closely related individuals.

If one analyzed four grapevines for several microsatellites, each individual would more than likely have a unique pattern, which is why the

ALLELES
#1 ——————— CACACACACACACACACACACACA CACACA —————
#2 ——————— CACACACACACACACACACACA CACACACACA —————
#3 ——————— CACACACA CACACACACACACACA CACACACACACA —————

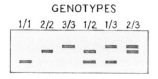

GENOTYPES
1/1 2/2 3/3 1/2 1/3 2/3

The dynamics of a microsatellite system. Microsatellite #1 has fifteen CA repeats, #2 has seventeen, and #3 has nineteen. Because there are three different sizes, when the fragments are separated in a gel, three different bands will appear, one band for each size fragment (fifteen, seventeen, and nineteen repeats). When plants from a population are assayed in a gel that separates the fragments by size, six different genotypes can exist, as shown.

microsatellite patterns are called fingerprints. Although this is an over-simplification of how microsatellite analyses are actually done, by simply counting the bands in common among the various pairs of individuals researchers will often be able to determine which individuals are close relatives and which are not.

A problem with microsatellites is that they need to be chosen randomly; and to make sure that there is no bias in the inferences made from their analysis, they need to be physically far apart from each other on the DNA strand (the chromosome). The best way to ensure this is to use microsatellites from different chromosomes. If two microsatellites are used the probability of getting unique stripe patterns for any four individuals is low, and for the entire population or species it would be impossible. But if five are used, the probability that each individual has a unique stripe pattern, or fingerprint, increases. Mathematically speaking, it takes only thirteen microsatellites to determine a unique genetic profile for any human being, and thus to identify a human individual. Scientists at the University of California, Davis, used grape microsatellites in the late 1990s to pin down the origin of Chardonnay, Gamay Noir, and other French cultivars. These early microsatellite studies demonstrated that these cultivars from northeastern France derived from offspring of matings between Pinot (in many

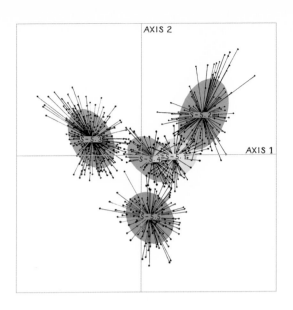

Multivariate analysis of different grape strains. The different geographic locations of the vines are shade-coded as indicated in the illustration of the phylogeny of wild grape strains. Note that there are five clusters corresponding to the five colored geographic regions where the vines were collected. Redrawn and modified from Myles et al., "Genetic Structure and Domestication History of the Grape."

cases Pinot Noir) grapevines and the humble Gouais Blanc, planted widely in the same region in medieval times but now widely shunned.

The use of microsatellites as indicators of relatedness expanded after the *Vitis* genome was sequenced in 2011. The genome sequence allowed Laucou and colleagues to characterize a larger number of microsatellites, chosen so that there would be at least one on each of the nineteen grape chromosomes. Of the 4,370 varieties analyzed, a little more than half turned out to have unique profiles, meaning that in microsatellite terms some of the cultivars and varieties were effectively identical, leaving only around 2,800 unique profiles. Using these latter profiles, the team could determine the relatedness of the different cultivars they had identified. They used a statistical approach that allowed them to plot each individual on a graph, in order to visualize the relatedness of the cultivars involved.

The more alike their microsatellite profiles, the closer two strains lay in the graph space, revealing how closely related they were. Two important findings emerged. First, the authors could visualize genetic differences useful in identifying each of the four major grape variety categories examined (cultivars, wild, hybrid, and rootstock). Second, the broad set of cultivars showed about the same variability as many other crop and forestry plants. With the first set of results in hand, the researchers could rapidly characterize any grapevine as a cultivar, rootstock, hybrid, or wild strain. And the second observation suggested that grape growers over the ages have not created a domestic plant lacking in genetic variation. This is great news for grape growers because any crop or species with low levels of genetic variation is more prone to loss or extinction.

The final recent study relevant to the origin of wine grapes comes from the United States. A team led by Sean Myles, then at Cornell University, and colleagues took a slightly different genetic approach. Because the *Vitis vinifera* genome has been sequenced, it can be scanned for places where there are differences between cultivars. These usually involve single nucleotide changes (hence they are called "single nucleotide polymorphisms" or SNPs), and any individual in a grape population can be studied for whether a G, A, T, or C exists in a specific position in its genome. The study is conducted using a microarray, an amazing miniaturized laboratory placed on a chip slightly bigger than a half-dollar.

In the procedure used by Myles and his colleagues, DNA from the cultivar, rootstock, wild strain, or hybrid is isolated from the leaves of the vine. The DNA is chopped into small pieces using high-frequency sound, and a fluorescent molecule is connected to the end of each of the sheared fragments. This single-stranded DNA is called the target DNA, and the microarray chip is used to determine which single nucleotide polymorphism states (G, A, T, or C) it contains. DNA does not like being single-stranded, so each of the fragments from the treated target DNA seeks out the best sequence on the microarray and sticks to it. In most cases, each DNA fragment will find its direct complement, and hence will divulge its single nucleotide polymorphism state by showing a fluorescent spot at the appropriate position on the microarray. This approach, also known as DNA resequencing, is a rapid way to generate large amounts of DNA sequence for numerous specimens.

Myles and colleagues used about nine thousand SNPs, covering all

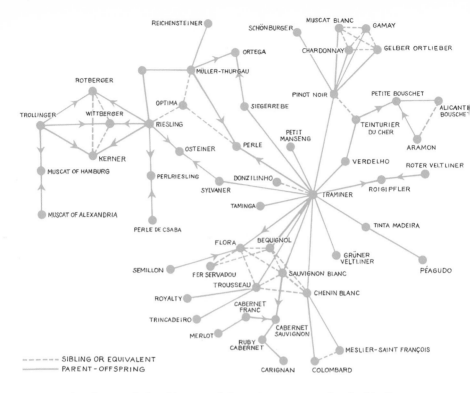

First-degree relationship network for various grape strains. In this diagram, strains that appear to be related to each other via a common parent strain are joined by lines. Redrawn and modified from Myles et al., "Genetic Structure and Domestication History of the Grape."

nineteen grape chromosomes. In an initial analysis using a chip affectionately known as Vitis9kSNP, the researchers analyzed about 950 cultivars (451 table grape accessions, 469 wine grape accessions, and 30 of unknown type) and 59 wild *Vitis vinifera sylvestris* varieties. Using the SNPs identified through sequencing of the *Vitis* genome, they conducted a parentage analysis to determine whether members of particular cultivar pairs were more closely related to each other than to others in the data set. What they were seeking was first-degree relationships potentially between parent and offspring. And their analysis produced some surprising results.

The data showed, first, that almost 75 percent of the varieties and cultivars examined had at least one first-degree relationship with another variety or cultivar, indicating a high degree of interconnectedness among the grape cultivars. Some had more than one first-degree relationship, which made tracking their genealogy difficult. But even when there was more than one (indeed, up to seventeen) first-degree relationships it indicated that the variety had been repeatedly used in the domestication of grapes—a result that satisfyingly confirms earlier scientific findings, as well as oral grape-breeding tradition. The researchers could, for example, detect seventeen first-degree relationships among Pinot and Gouais Blanc grapes, close to the sixteen suggested by historical studies in the 1990s. Similarly, Chardonnay samples showed seven first-degree genetic relationships, corresponding exactly to relationships documented by early microsatellite studies. Most of the grapes with large numbers of first-degree relationships were table grapes, which may imply that table grapes have been more inbred than wine grapes.

The team could also infer with some confidence that half the first-degree relationships observed were probably parent-offspring relationships, whereas the other half were "sibling or equivalent," which is half as close as first-degree relationships. They provided a table of all 950 cultivars they analyzed, showing the most likely first-degree and sibling or equivalent relationships. In some cases, the origin of a particular grape strain was known from written records or from oral tradition, though often, even when documentary and genetic information could be combined, the origins of many wine grapes remained hazy. But perhaps most important, what Myles and his colleagues have achieved is to give grape growers a veritable studbook for understanding the place of origin of certain grape varieties.

Some notable connections they made from the genetic data are that "Chenin Blanc and Sauvignon Blanc are likely siblings, and both share a parent-offspring relationship with Traminer . . . [and] . . . two of the most common cultivars of the Rhône Valley in France, Viognier and Syrah, are likely siblings." The second observation is particularly useful, because one of those two grape cultivars produces white wine and the other red. Myles and colleagues neatly summarize their results on cultivar relationships by noting, "These observations suggest that grape breeding has been restricted to a relatively small number of cultivars and that only a small number of the possible genetic combinations within *vinifera* have been explored." There is a lot left to do.

Although the overall goal of this research was to characterize all the cultivars and accessions in the USDA grapevine collection, the team also looked into the wild progenitor question. Using the same graphing approach favored by the Laucou group, Myles and colleagues showed clearly that all the grape cultivars they looked at ultimately come from the Near East, as broadly defined—so Areni's claim to be the birthplace of wine remains intact, at least on this basis. Still, when they branched out and examined wild grapes (*Vitis vinifera sylvestris*) from Armenia, Azerbaijan, Dagestan, Georgia, Pakistan, Turkmenistan, and Turkey, they were unable to pinpoint the location of the progenitor of all cultivated grapes—most likely because there has not been enough differentiation among western Asian wild strains over the past few thousand years for them to be accurately characterized.

Approaches using genetic analysis have thus helped greatly in unraveling the relationships among grape cultivars. Recent studies of seed anatomy have also produced results that are generally in agreement with the molecular research. (Scientists always like it when different kinds of data sets converge.) By combining comparisons of seed morphology in wild and cultivated strains from the Vassal collection, Jean-Frédéric Terral and colleagues came to some striking conclusions about the origin of wine grape cultivars, using a statistical approach similar to the one used for microsatellites. The closer two strains lay in the graph space, the more alike their seeds were and, so the reasoning went, the more closely related they were. The seed shape of the wild strains analyzed indicated a very close relationship to the modern Clairette cultivar, a rather obscure grape most widely grown today in southern areas of France. More detailed analysis also revealed a connection between the equally obscure Mondeuse Blanche cultivar (today grown almost exclusively in the Savoie region of eastern France) and wild grapevine strains. One visually appealing way to represent these data is by using a branching diagram such as the one depicted in the illustration, in which many of the well-recognized groups of today's cultivated grapes still "hang together." Once again, the

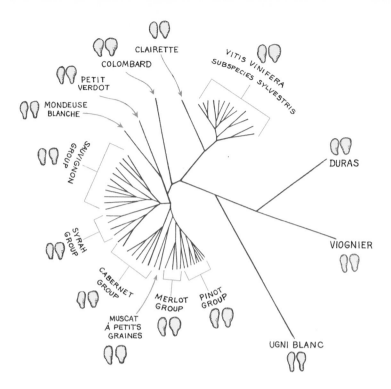

Family tree for grapes based on seed shape. Seeds of different strains were collected and analyzed for size and shape. These data were then transformed into similarity measures: the tree here represents the similarities among the seeds of different strains. Redrawn and modified from Terral et al., "Evolution and History of Grapevine (*Vitis vinifera*) Under Domestication."

Clairette cultivar appears to be most closely related to the wild strains, and Terral and his colleagues argued, "If the existence of this variety is confirmed by new data, such as from current archaeogenetic investigations, 'Clairette' would become one of the oldest authenticated varieties."

Terral's team followed up on its own suggestion by examining well-preserved archaeological seeds, using seed shape as a "fingerprint" to indicate which cultivated or wild strains the archaeological specimens most resemble. They looked at fifty ancient seeds from archaeological sites near the southern French city of Montpellier dating between 75 and 150 c.e. and found they could classify thirty-four. Among these, they identified ten wild grape seeds, eight seeds of the Merlot group, six of the Clairette group, six of the Mondeuse Blanche group, and two each of the Cabernet Franc and the Hanab-Muscat groups. Evidently, not long after the beginning of the Common Era, the ancient inhabitants of southern France had already explored using not only the Clairette and Mondeuse Blanche strains but also several others that are more familiar today. Finally, by comparing grape seeds currently growing in the Languedoc region of southern France with grape seeds found at archeological sites there, the group concluded the Languedoc was a center of intense domestication of wine grapes in the centuries preceding about two thousand years ago, a finding that supports the historical record.

5

Yeasty Feasts

Wine and Microbes

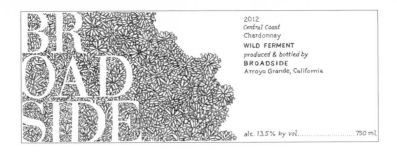

2012
Central Coast
Chardonnay
WILD FERMENT
produced & bottled by
BROADSIDE
Arroyo Grande, California

alc. 13.5% by vol........................ 750 ml

"**W**ild Ferment" said the label on the bottle. So, well aware of the recent Internet debate over whether wild yeast fermentation actually exists, we were intrigued about the wine we would find inside. At least, we thought, with no added commercial yeast and minimal intervention in the winery, this wine should impart a sense of terroir (the spirit of the place) too often lacking in California Chardonnays in its modest price range. And though we were well aware that our taste test was an experiment without a control, our expectation was fulfilled: the wine was refreshingly lean and acidic, avoiding the overblown fruitiness of many of its counterparts.

Through studies such as the ones we described in Chapter 4, scientists are learning a lot more than they once thought possible about the origins of grape cultivars and their relationships to one another. This endeavor is important, because the history of domesticated grapes tells us much about the history of wine. And in the future, knowing which cultivars are most closely related will be important in the breeding of vine stocks.

But much of the chemical complexity in wine is produced not simply by the grapes but by their partner, the yeast. And it is to this remarkable organism that we now turn. Yeasts have a less convoluted evolutionary past than grapes, but their story is equally absorbing. In fact, many of the questions we just asked about grapes have also been asked of yeast, and have been at least partially answered. These include "Which wild yeast strain is the progenitor of the yeasts that are used in wine making?" and "Where did this 'mother of wine yeasts' originate?"

Yeasts are fungi. But unlike the more familiar mushrooms, they are not as easily characterized by their morphology as mushrooms are, largely because of their nondescript anatomy—which, because they are incredibly tiny, needs to be viewed through a microscope. The yeasts involved in winemaking come primarily from a single family, known cumbersomely as the Saccharomycetaceae. This family contains thousands of species, but one in particular, *Saccharomyces cerevisiae,* is essential to wine production. This yeast species reproduces both sexually and asexually. When reproducing asexually, each cell "buds" to create rather bloblike daughter cells, to which a duplicated nucleus is then transferred.

Budding yeasts are fairly common in the environment, wherever ample carbohydrate sources like sugar are present. Unlike plants that can use both nutrients and sunlight to produce energy, fungi need nutrients such as carbohydrates for this purpose. Yeasts are nonetheless abundant, and because many of them can easily be grown in the laboratory, they have been the subject of close scientific study—appropriately enough, because the yeast species that is generally used in winemaking (and in baking bread and brewing beer) is one of the most important domestic species on this planet. Scientists love *Saccharomyces cerevisiae* because it is a fantastic model genetic organism. It grows fast; it is easy to cultivate in the laboratory; and it is, of course, a eukaryote, as are grapes and humans. For all these reasons, it is a useful form with which to study how proteins interact, and how genes are involved in those interactions.

When whole-genome sequencing came along in the 1990s, *Saccharomyces cerevisiae* was an obvious candidate for sequencing. In fact, it holds the distinction of being the first eukaryote to have had its entire genome characterized, in 1996, on the heels of the first free-living organism, a bacterium called *Haemophilus influenzae.* Subsequently, almost two dozen other species in this yeast family have also had their genomes sequenced.

Because fungi are single-celled organisms, they might at first seem pretty boring. But when we examine the various lifestyles that even such simple creatures can adopt, a stunning array of species and evolutionary patterns emerges. To illustrate this phenomenon, we need look no farther than our own everyday lives. Hardly a day passes in which we don't eat a food produced using a fungal species. (Sometimes the food itself is fungal, such as mushrooms and truffles.) Fungi might also have caused some of our most uncomfortable illnesses, as well as many minor ailments such as athlete's foot. For some, fungi might even have been the source of mind-expanding experiences: psilocybin compounds found in over 150 species of fungi are famous for their psychedelic effects.

✦ ✦ ✦

Our current understanding of the fungal tree of life was developed by a large collaborative group of researchers led by Rytas Vilgalys at Duke University. These scientists focused on six genes for each fungal species they examined, and they used the DNA sequences of those genes to construct a genealogical tree for about two hundred of the better-known species of fungus. This tree confirmed much of what was already known about fungal relationships, but it also indicated the position of several new groups for the first time. Still, this was just a start. There are by now around one hundred thousand formally described species of fungi, and some researchers suggest that there may be between 1.5 million and 5 million species. If this number seems improbable, bear in mind that, although seven thousand bacterial species have been described, many scientists think that there may be 10 million to 100 million species of them!

The fungal tree of life tells us that there are two major kinds of fungi, along with several "lonely" groups. These latter deserve our attention as novel and separate, but they contain few representatives. The two major kinds of fungi are the Basidiomycota (puffballs, mushrooms, and stinkhorns, for example) and the Ascomycota, the group within which the fungi important to wine are found. They differ fundamentally in reproductive style. And although the major player in the making of wine is the ascomycete *Saccharomyces cerevisiae,* there are several other fungi that also influence wine production, both positively and negatively. The table lists the dirty dozen of winemaking, giving the classification of the fungal species with which the winemaker has to be concerned: those that come into play at one time or another during the fermentation process. Note that most are ascomycetes belonging to the family Saccharomycetaceae. But a couple of basidiomycetes also serve a function in winemaking.

As with the grapes, we will not have finished telling the story of yeast until we have addressed the origins of the strains used in wine produc-

Phylum	Order	Family	Genus
Ascomycota	Saccharomycetales	Saccharomycetaceae	Hanseniaspora
Ascomycota	Saccharomycetales	Saccharomycetaceae	Saccharomyces
Ascomycota	Saccharomycetales	Saccharomycetaceae	Candida
Ascomycota	Saccharomycetales	Saccharomycetaceae	Pichia
Ascomycota	Saccharomycetales	Saccharomycetaceae	Kluyveromyces
Ascomycota	Saccharomycetales	Saccharomycetaceae	Torulaspora
Ascomycota	Saccharomycetales	Saccharomycetaceae	Brettanomyces
Ascomycota	Saccharomycetales	Saccharomycetaceae	Dekkera
Ascomycota	Saccharomycetales	Saccharomycodaceae	Kloeckera
Ascomycota	Saccharomycetales	Metschnikowiaceae	Metschnikowia
Basidiomycota	Tremellales	Tremellaceae	Cryptococcus
Basidiomycota	Sporidiales	Sporidiobolaceae	Rhodotorula

tion. By 2005, only a handful of yeast species had been examined at the genetic level for whole-genome variation. Today a yeast genome can be sequenced in less than a day, and for a fraction of what the first yeast genome sequence cost, so researchers have now analyzed several hundred yeast strains in their quest for the closest wild relative to the yeast strains essential in making bread and wine. The scientists who work on this problem call these domestic strains "captive yeasts." An aid in this quest has been the existence of centralized repositories for *Saccharomyces cerevisiae* strains and their close relatives. One of the largest of these is in Great Britain, at the Institute of Food Resources in Norwich, which contains more than four thousand strains.

The approach to determining a yeast progenitor is similar to that used to find a grapevine ancestor, with the closest wild species and subspecies used to anchor the search. In the case of yeast, this started with the characterization of a yeast species called *Saccharomyces paradoxus*, chosen because it has "escaped captivity" and can serve as a baseline for what *S. cerevisiae* might have been like if it had not been "captured." Then, against a background of yeast strains from a variety of origins—vineyards, sake factories, medical samples, and natural sources such as in fruit or tree exudates—researchers examined the population structure of wine yeast strains. Their most notable finding was that sake yeasts and wine yeasts show a well-defined separation, indicating that they have been kept sepa-

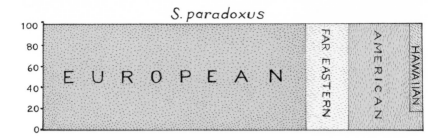

Genetic structure of the yeast Saccharomyces paradoxus. Individual strains are represented by columns. The diagram shades the different yeast strains by the localities with which they can be associated. For instance, the left-most strain can be designated 100 percent European, and the strain on the far right is about 80 percent Hawaiian and 20 percent American. Genetic structure for these strains correlates strongly with geography. Redrawn and modified from Liti et al., "Population Genomics of Domestic and Wild Yeasts."

rate since each was first used to ferment the relevant beverage. This suggests that two distinct instances of the application of human ingenuity (or luck) resulted in the capturing of the yeast strain involved.

Using a larger sample and whole-genome sequencing, a group from Great Britain increased the resolution of their analysis. And even though Saccharomyces paradoxus and S. cerevisiae have similar ecological preferences, they turned out to show some significant differences when their genomes were examined over wide geographic ranges. Having established which genes of the wild population were typical of which geographic areas, the British scientists used an approach called STRUCTURE analysis to determine how "mixed" various populations were. They assigned each geographical gene a different color and then looked at the color spectrum characteristic of each local yeast population. If, for example, the S. paradoxus genes that best represent Europe were given the color blue, Asian ones yellow, and American red, any unambiguously European individual would be represented entirely by blue, whereas individuals of mixed origins would also have yellow and/or red. S. paradoxus, the researchers discovered, showed a large degree of solid coloring in STRUCTURE analyses, indicating that there has been little genetic contact and mixing among geographic regions. Geneticists call this a "highly structured" pattern.

In contrast, Saccharomyces cerevisiae displayed an unstructured pattern

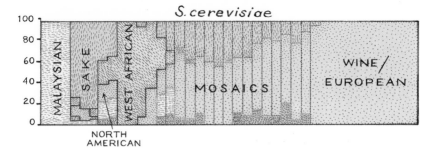

Genetic structure of the yeast *Saccharomyces cerevisiae*. Individual strains are represented by columns. The diagram shades the different yeast strains by the localities with which they can be associated. For instance, strains in the middle of the diagram have multiple probable geographic origins, whereas the strains on the far right can be designated as 100 percent European (wine) and those on the far left as Sake, North American, and Malaysian. In the middle of the diagram there is little genetic structure based on geography among the strains. Redrawn and modified from Liti et al., "Population Genomics of Domestic and Wild Yeasts."

that revealed extensive mixing. This multicolored pattern indicates a great deal of nonnatural genetic manipulation, which is what we would expect from captive organisms such as the domesticated yeasts. The same analysis also revealed that once a yeast strain was established for winemaking, it became strongly homogenized. One conclusion: Don't try to improve on what has happened naturally—if a yeast is making good wine, don't outbreed it.

The British researchers also took a closer look at the question of where the "mother of wine yeast" originated. Using genome-level sequence information, they were able to generate a genealogy for the yeast strains used in winemaking. This genealogy confirms the separation of sake yeasts and wine yeasts inferred earlier from fewer genes, but it also shows the close associations of all kinds of captured yeasts with the yeasts used to make wine. Evidently, not just one or two domestication events were involved, and this mishmash of a genealogy indicates a complex human influence in crossbreeding among strains.

◆ ◆ ◆

Still, *Saccharomyces cerevisiae* is not simply a human plaything. Samples preserved in amber show that this yeast species existed as much as 30 million years ago, and it was presumably already fermenting ripe fruit

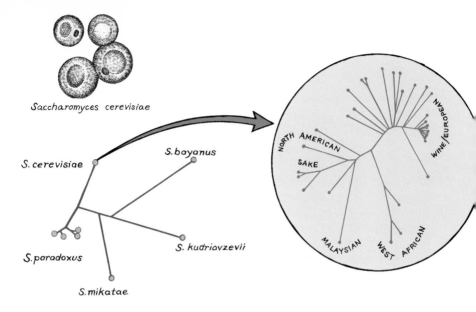

Wine yeast phylogeny. The diagram at the lower left represents the relationships of the yeast species most closely related to *Saccharomyces cerevisiae*. The larger tree on the right is an expansion of the *S. cerevisiae* part of the tree. Redrawn and modified from Liti et al., "Population Genomics of Domestic and Wild Yeasts."

long before humans came along and coopted it to their purposes. What is more, humans are not the only animal species with which S. *cerevisiae* has an intimate reciprocal relationship. Tiny as they are, these microorganisms do not simply float around in the air, waiting to alight on a convenient bunch of grapes. They have to be moved by an animal vector. In 2012 an ingenious genomic study carried out in Italy by Duccio Cavalieri and his colleagues fingered the predatory wasp *Vespa crabro* (the European hornet) as a crucial player in the life cycle of this yeast.

It has long been known that S. *cerevisiae* colonizes the developing grape bunches anew each spring and does not overwinter on the vines themselves. So where were the yeasts when they weren't on the grapes, and how were they getting to them? Cavalieri and his team showed that the intestine of *Vespa crabro* periodically shelters several different yeast species, but only S. *cerevisiae* can always be found there. Analyses using microsatellites and a variety of gene loci further showed that local populations of wasps harbor typical strains of S. *cerevisiae*, with which they clearly have an enduring mutual relationship. The yeasts are transferred from one generation of V. *crabro* to the next when the adult wasps feed their larvae by regurgitating insect prey already digesting in their intestines. Once the larvae have metamorphosed into adults and can fly around in search of food, they use their strong mouth apparatus to puncture the tough grape skins to extract the sugars within. As they do this, they give the grapes a natural inoculation of the winemaker's preferred type of yeast. And the yeasts themselves are given a flying start in the business of fermentation, though this may not always be to the winemaker's delight.

We aren't finished quite yet with the various organisms involved in winemaking, because the total number of species involved is so large that the best approach is to consider all as members of an ecological community, each one diligently doing its own job in making wine. In the next chapter we'll examine the complex ecological dance in which they participate along with many other organisms. But here we'll take notice of the actors involved in general terms, even if we can't yet be sure of the roles each of them plays in vinifying the grape.

If you were to take a single grape from the vine, or even a spoonful of the soil the vine grows in, you would find millions of organisms. Some, such as the nematodes in the soil, would be rather large; but there are also plenty of extremely tiny, technically nonliving entities called viruses. Most of the life in the spoonful of dirt would be microbial, composed of some fungi, some archaeans, some protists, and lots of bacteria. Why are all these microbes there? What are they doing, and how can we observe them? Robert Tiedje, one of the founders of modern microbial ecology, once quoted a London student's description of a soil community of microbes, likening it to a city: "On a human-eyed scale, the soil for a bacterium must be like living in a 30 km-high, crumbling, dark Bladerunner-esque city that is often deluged with water, packed with garbage, and full of all manner of modest and badly ventilated dwellings. Apart from the extra dimension, pretty much like London in winter, in fact. The only difference with our fine city is that I suppose the landscape would be peppered with catabolic fires around root tips, within and in the wake of worms, and following the death of roots and soil organisms creating spectacular opportunities for several trophic groups."

This is a great metaphor for the interactions that are going on around us all the time at the microscopic level. The living world is saturated with interactions among the species that constitute it, and neither we nor any other large organism could function without all the life going on, unrecognized, within us. Over 90 percent of all the DNA that is found in our bodies is not our own but comes instead from microbes—and we would be in big trouble without them.

6

Interactions

Ecology in the Vineyard and the Winery

One recent early summer day, we had the pleasure of visiting several wineries in northern California's Alexander Valley, one of the state's most outstanding growing regions. The sun was shining brightly; the sky was clear; and a warm wind swept over the valley. We sat close to the edge of the vine rows, sampling a glorious local Cabernet Sauvignon and watching as the light breeze ruffled the glinting leaves on the receding rows of vines. As lazy observers we appreciated the pastoral beauty of the scene. But what the biologists in us also saw was a landscape of sex and death.

The fields of ecology and evolution could be characterized as studies of sex and death in nature. And although both fields are fairly new as scientific disciplines, the spirit underlying them is age old. Both Aristotle and Hippocrates wrote descriptive accounts of the natural world around them, and it escaped neither that the objects through which they described that world were knit together, and given meaning, by the interactions among them. Aristotle's fourth-century B.C.E. protégé Theophrastus wrote extensively about plants, and was specific about how they should be understood. In the prescient statement that opens his classic *Enquiry into Plants* he made his approach clear: "We must consider the distinctive characters and the general nature of plants from the point of view of their morphology, their behavior under external conditions, their mode of generation, and the whole course of their life." Theophrastus was, in fact, describing an ecological evolutionary approach to understanding plants, and ultimately he extended this focus on interactions to grapes and other fruits in a short work titled *On Wine and Olive Oil*. There he discussed the ripening of fruit (especially wine grapes) in the specific context of environmental conditions, particularly sunshine and heat.

So why have biologists been so obsessed with sex and death in the natural world? No mystery there—that's where much of the action is in the ecological and evolutionary spheres. Evolutionary biologists often use the term "life history" to denote the ways organisms have evolved to be reproductively successful. And each life history strategy they recognize carries with it a reference to the potential contribution each individual will make to the next generation of its population and species. This is a major part of the evolutionary dynamic, and although it is not the whole story of change in the living world, it is omnipresent. A cultivated grapevine goes through several stages on its way to making grapes: bud break, flowering, fruit set, véraison, harvesting, leaf fall, dormancy. This life cycle is a somewhat artificial one, because it differs significantly from the natural cycle (in which there is no harvesting: the grapes interact instead with a variety of frugivores). But the vine remains in rhythm with the seasons.

A life history, however, is not just the story of an organism's development, or even of its life cycle. Instead, a life history encompasses all the traits that are important for the organism's reproduction and individual survival. Age at first reproduction, fecundity, and age at last reproduction are all important to biologists studying the evolutionary success of species. So how do grapevines manage, and how have their life history strategies contributed to their success?

✦ ✦ ✦

One of the most important evolutionary problems any organism needs to solve, whether it's a bacterium or an elephant, is how to create the next generation. Even viruses, not usually considered living organisms, are prolific and rapid reproducers. Prions—proteins that do not even have a genome—also contrive to replicate themselves copiously. And in the same spirit, the need to reproduce has shaped various parts of the grapevine. These plants spend most of their energy budget making leaves, seeds, rootstocks, and fruit. Why? Well, leaves and rootstocks are easy to understand. They are essential for the vine's maintenance, converting the sun's rays to useful energy and transporting nutrients throughout the plant. But how about the key component, the grapes that hold the seeds? They, too, represent a huge energetic investment, but they are on the vine for a different reason.

Reproduction is pretty straightforward for organisms that can walk, crawl, or slither in the interests of spreading their gametes around. Such creatures move hither and yon, hoping to find another with which to reproduce. But a plant can't do that. Male plants have solved half of the problem posed by their immobility by packaging their gametes into tiny, light particles called pollen, and have come up with many ways to ensure that the pollen is spread around. The most spectacular of these involve coopting organisms that *can* move, analogous to yeast's exploitation of wasps. Some plants are thus experts at attracting unsuspecting insects to do the heavy lifting. Others have taken a different tack, dispersing their seeds over long distances by means of tricks that allow their gametes to float on the air. Yet others have evolved a carnival sideshow–like half-woman/half-man strategy. Grapevines are among this last group.

The importance of such mechanisms is reflected in their sheer ingenuity. Charles Darwin put it this way in *On the Various Contrivances by Which British and Foreign Orchids Are Fertilised by Insects* (1862): "An examination of their many and beautiful contrivances will exalt the whole vegetable kingdom, in most persons' estimation." He went on to explain the contrivances of what he clearly saw as the most ingenious orchid of all, the genus *Catasetum*. Also known as "Darwin's bee trap," this orchid has a hair-trigger organ near the entry of the male flower. When tripped by a bee, the trigger shoots out a dart with pollen attached to its end, at a speed of more than 300 centimeters per second. The dart, a "pollinium," sticks to the back of the bee, which will then deliver the pollen to the ova of a female *Catasetum* on its next visit. Domesticated grapevines keep things simpler, since most of them are hermaphroditic and are thus able to pollinate themselves. Insects and wind only rarely intervene.

Yet even when they can accomplish fertilization unaided, grapevines still need help in dispersing their seeds. Plant seeds can be scattered in several ways. One is by simply adhering to an animal that passes by. Anyone who has tramped through a field of high grass knows that you're likely to come away with burrs sticking to your socks or pants. Those seed-containing burrs will either fall off or be picked off, and they'll probably wind up on the ground some distance away from the parent, where they can carry out the business of reproduction. Another plant trick is dispersing the seeds through the air. Dandelions do this by attaching their fertilized seeds to a fluffy apparatus that can float in the breeze; maple trees have evolved a helicopter-like mechanism to disperse their seeds; and tumbleweeds have invented the wheel.

Still, perhaps the most popular long-distance dispersal strategy plants use is having their seed-containing parts eaten by animals. This strategy has dictated three of the more important aspects of grape anatomy—their heavily coated seeds, which can withstand the rigors of the stomach acids and intestinal enzymes to which they will be subjected before they are excreted; their color, which will catch the eye of potential dispersers; and their sugary innards, which given them an appealing taste. Thus, the biochemistry that makes grapes sweet came about because the vine evolved to bear fruit that would be attractive to potential dispersers. And it's a competitive world out there—lots of plants are designed to disperse their seeds this way, usually at around the same time. So the fruit of the vine needs to be both as eye-catching and as tasty as possible.

This may explain why grapes are often red. Recent experiments suggest that birds, at least, greatly prefer red things to blue, yellow, green, or black ones. Researchers have demonstrated this preference by removing newly hatched birds from their nests, raising them in isolation, and then offering them items of different colors. The birds showed a clear preference for red objects. Vines generate the red color found in their skins and pulp by producing anthocyanins, rather bulky molecules that belong in the flavonoid group. So effective are these pigments that they are used in the food industry to produce food colorings, and they have the possible additional advantage of being antioxidants, often touted by health advocates as helping to counteract damage to the tissues caused by electron-robbing free radicals. (There is now a medium-sized question mark hovering above the potential health benefits claimed for antioxidants as a group, but there are compounds in red wine—notably the phenol known as resveratrol—that may be associated with some degree of cardiovascular benefit.)

Scientists have understood the molecular pathway involved in producing anthocyanins for a long time. The chain of reactions involves many proteins, each of which has a specific job in molding the structure of the anthocyanin. In the past decade, researchers have also begun studying the genes controlling all this activity. Whenever a gene is responsible for a physical trait (such as red color in grapes), one of two processes can occur. The gene concerned can make a protein that has a direct, physical impact on the trait, in which case it is called a "structural gene." Or it can act like a valve, regulating the production of the protein involved. Genes of the second type are called "regulators," or "transcription factors," because they regulate the amount of protein produced, as well as when and where it is made. By studying entire grape genomes, researchers in Portugal have determined that there are ten structural genes and five regulators involved in grape coloration, a complexity that explains the extraordinary range of tints and color densities found in wine.

One of the more surprising discoveries of recent studies is that structural genes tend not to vary much among related organisms, and are sometimes not even active in producing the traits with which they are usually associated. Rather, it is the regulatory genes that do a lot of the dirty work, and implement much of the variety we see in nature. Grape color is no exception. The major genes controlling the production of anthocyanins in grapes are called Myb, Myc, and WD40. Japanese investigators have shown that the light color of Koshu "pink grapes" is caused by lowered anthocyanin production due to a defective Myb gene. The Koshu Myb gene has two small extra domains (the 44 and 111 nucleotides of DNA) in one of the Myb genes, an alteration that apparently reduces the amount of anthocyanin in the grape skin.

As we have already mentioned, nobody has yet been able to produce a seedless grape that makes good wine, so the seeds buried within the grape must provide an essential element of the chemical complexity of concentrated red wines. But grape seeds nonetheless pose a bit of a dilemma for the winemaker. On the one hand, they are critical to the reproductive success of the wild vine, and their presence is associated with good wine. But they are also the waste products of the winery. Grape berries evolved to attract animals, who would eat them and disperse the seeds; but if the

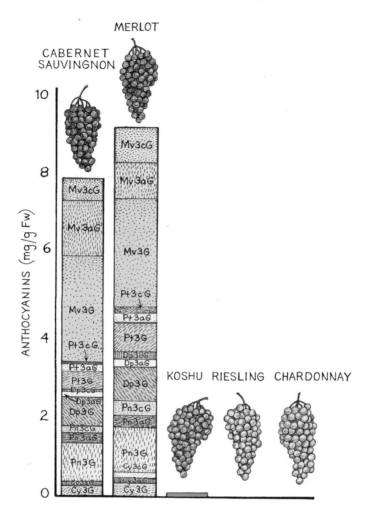

Anthocyanin content analysis of Cabernet Sauvignon and Merlot. The amount and types of anthocyanins are shown in the bars in the diagram. Also shown on the graph are Koshu, Riesling, and Chardonnay; the two latter have insignificant anthocyanins, while Koshu has very few. "Mv," "Pt," "Dp," "Pn," and "Cy" are abbreviations for the various kinds of anthocyanins found in these wines. Redrawn and modified from Shimazaki et al., "Pink-colored Grape Berry Is the Result of Short Insertion in Intron of Color Regulatory Gene."

seeds are digested, they can't be dispersed. It is thus necessary for the animal to excrete them intact, after carrying them away from the vine. Hence the seeds of grapes, and indeed those of any other plant using this mode of dispersal, have acquired hard coats to help them survive the journey through the masticatory and digestive tracts of large animals. These coats make grape seeds tough, but as a second line of defense they also contain some nasty chemicals that an animal consuming the fruit would prefer not to taste. And this, of course, poses a problem for winemakers.

The grape seed, as we saw in Chapter 4, is composed of an outer seed coat, an endosperm, and the embryo nestled in the center. The seed coat acts as armor during the hazardous journey through the digestive tract, while the embryo is the seed's precious cargo. The intermediate endosperm is fleshy, and provides nourishment for the embryo until the seed decides to germinate. These three layers all contain unpleasant-tasting polyphenols, which can make up almost 10 percent of the seed's volume. But many other compounds are present, too, some of them noxious, though not all. Indeed, grape seed extract is controversially promoted for a range of health advantages, and grape seed oil is an excellent frying medium because it will not burn until it reaches very high temperatures. But from the wine drinker's point of view the key thing is that, for whatever reason, clarets, Chateauneufs, and Cabs would just not be the same without the seeds.

✦ ✦ ✦

The interactions of grapes with the dispersers that eat them are easy to observe. This is why most of the initial thinking done about interactions among organisms was about animals and plants that are visible to the naked eye. But it is now some time since scientists began to discover the role of microbial life in infectious disease. Diseases such as childbed fever and other killers have decimated human populations since time immemorial, and studying the microbes that caused disease was an important step in advancing human health. But this is only part of the story. Around the turn of the twentieth century two microbiologists, Martinus Beijerinck and Sergei Winogradsky, realized that microbes were everywhere and affected many natural processes; they did not simply cause diseases. Winogradsky was the first to realize that microbes were responsible for the enrichment of soil with nitrogen, while Beijerinck was one of the first scientists to obtain cultures of agriculturally important bacteria and bacteria involved in plant ecosystems. For years, researchers interested in this fledgling science kept their day jobs, so to speak, and studied microbial ecology as a side interest. But the more that has been learned, the more important the microbiome has become to our understanding of its role in enabling the conduct of our daily lives. Each of us is crammed with microbes that are essential for innumerable bodily processes.

The major difficulty in studying microbes is that they can be seen only with powerful microscopes. And until the past decade, the amount of DNA and protein in a single bacterial cell was too small for microbial ecologists to study using the techniques at hand. These scientists had to resort to culturing microbes from environmental samples, and what they studied was whatever they could grow in a laboratory. If, for example, researchers wanted to study the microbes living on a grape, they would wash the grape in a solution of water or dilute salt, take the wash, and try to culture microbes from it. The problem was that many microbial species could not be cultured. In fact, even now scientists haven't figured out how to culture between 95 and 98 percent of all bacterial species. As a result, other methods had to be invented to look at the microbes involved in the ecology of the grape.

One new approach capitalizes on the fact that DNA is not only a double-stranded molecule, but also a complementary one: a researcher with one strand of a double helix can figure out how the other is structured. Whenever there is a G (guanine) on one strand of the double helix, directly across from it on the other strand will be a C (cytosine). Likewise, if there is a T (thymine) on one strand, directly across from it will be an A (adenine). These stick together by means of chemical bonds (A with T, G with C). As we saw, DNA with the sequence GATCGATC on one strand will have CTAGCTAG on the other, and the Cs and Gs and the As and Ts will stick together, acting act like a zipper. If the double helical molecule is heated, it will start to unzip. And it will zip back up when it cools.

Now, imagine that our old friend the yeast *Saccharomyces cerevisiae* has a unique sequence in its genome such as GCATCATCGATCGAGCATG-ATCGCAGC. Somewhere in the yeast's genome the complement to this sequence exists on one strand. If this sequence is mixed with DNA from a yeast cell, heated, and then cooled, that sequence will find its complementary sequence and stick to it. Next, imagine putting a little fluorescent molecule on the end of this sequence, and repeating the exercise. What will happen this time is that, as expected, the sequence will stick to the complementary sequence in the yeast cell. And where it sticks, a tiny bit of fluorescence will be visible, indicating that the cell has the marker sequence, and hence is a yeast cell. If we have a number of sequences that we know are unique to a particular organism, we can make as many DNA probes as we wish and connect them to different-colored fluorescent beacons.

This approach is called fluorescent in situ hybridization, or FISH, and it allows us to know what species are present in a particular microscopic field and how many there are. FISH is used both for clinical purposes and to identify bacterial and other microbial species in samples taken from nature. This colorful way of identifying microbial species can tell scientists what kinds and quantities of microbes live on a grape—or anything else—and give them an idea of the players in the ecological game of wine.

If we were to use FISH on a grape sample, we might get an idea of the kinds of microbes living on the skin of the grape or in the dirt around it. But we would see only the species we have probes for. How then do we "see" all of the microbial species on and around a grape? Researchers learned in the 1990s that from a spoonful of dirt or a swab from the outside of a grape, they could make DNA much as a human genetics lab makes DNA from blood. The only difference is that in the DNA from blood there would be a single genome (the subject's), whereas in the spoonful of dirt or grape wash there would be the genomes of millions of microbes.

The mixture of DNA from the dirt or wash contains the genomes of all of the microbes in the spoonful. Since each piece of DNA in the sample comes from a particular species, the most logical procedure would be to sequence all the fragments of DNA in the sample. But prior to what scientists call next-generation sequencing (NGS: a total misnomer because we are seven years into "next" generation already), the process of obtaining DNA sequences was laborious and yielded little data for the effort expended. Only a small number of DNA sequences could be obtained from the sample—maybe between 500 and 10,000, a tiny proportion of the whole. These sequences could then be compared to a huge compendium of bacterial DNA sequence information, the ribosomal database (RDB). By matching the sequences from the sample with those from the RDB, a researcher would know which species had been sequenced in the dirt or grape wash.

Next-generation sequencing has upped the ante. Typically, it will yield between 400,000 and 10 million different microbe sequences. It also allows scientists to "see" the millions of microbes that can't be cultured in the lab. What this means is that researchers are getting more complete pictures of an increasingly large number of different microbial communities—and there seem to be an incalculable number of such communities in the world. It also means that new kinds of microbes are being discovered every day in media such as dirt, pond scum, air, seawater, sewage, and even on and in the human body. And NGS has given vineyard scientists an unprecedented perspective on life processes on the surface of grapes, on the soil in which grapevines grow, and in the grape must itself.

✦ ✦ ✦

The first big task of the grape microbial community biologist is to identify the species involved—basically, to do a broad census. Three major kinds of microbes are found living on the surface of grapes: filamentous fungi, yeasts, and bacteria. It is clear from microbial community studies that the particular species of all three kinds vary from strain to strain of grape; they also differ in presence or frequency from region to region. A study that counted microbes on Cabernet Sauvignon vines and berries showed that nearly half of the microbes on these plants consist of just ten major species, whereas the number of species found in a spoonful of dirt is generally more than a thousand. Equally intriguing is the finding that the community of microbes living on vine leaves is quite different from the community living on the grape skins.

Once the initial census is made, scientists can look at these communities in two major ways. First, they can ask what the differences are in the microbial communities living on different strains of grapes. This is important for an understanding of the extent to which a particular microbial community might be responsible for the characteristic taste of, say, a Cabernet Sauvignon. Second, researchers can ask how the microbial community changes on grapes of the same strain. The microbes on grapes

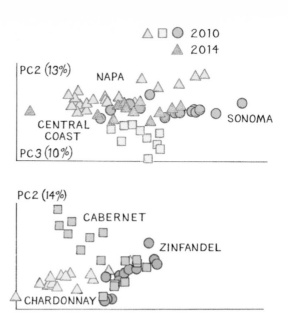

Graphs showing wine-growing region and grape variety correlated with bacterial community. The graphs were generated as multivariate statistical plots, where the axes become what are called principal components (PC) that explain the better part of the variation in the study. Note that the points cluster for the various shadings in both graphs, indicating that both geographic locality (*above*) and grape variety (*below*) are involved in determining the kind of microbes on the specific grapes. Redrawn and modified from Bokulich et al., "Microbial Biogeography of Wine Grapes Is Conditioned by Cultivar, Vintage, and Climate."

change as the berries ripen, and there is a huge turnover of the major kinds of bacteria living on grapes at the beginning of the ripening season, as well as a strong component based on environmental and varietal factors. In a 2014 study using NGS, Nicholas Bokulich and colleagues at University of California, Davis, examined the dynamics of the microbial makeup on grape skins in northern California. They were interested in what factors were involved in the makeup of the bacterial communities on grapes and in grape must. This is an important question for viticulturists because it is a first step toward circumventing the damage done to grapes by bacterial species. If the interactions and communities had proved to be random, it would have complicated attempts to remedy bacterial infection. But by identifying the microbial community structure on grapes from the Napa and Sonoma Valleys and in the central coast region, the scientists showed that bacterial communities are nonrandomly associated with grape must and vary by region, variety of grape, and other environmental factors.

Another study revealed that fungal species living on grapes often change depending on where the vine is in the vineyard, indicating a very fine spatial arrangement of microbes even within the same planting. Finally, it seems that the microbial communities in the same vineyard also vary from harvest to harvest, so clearly the interactions involved are incredibly complex. It remains for future research to explore the implications of these interactions for the production of wines having a particular desired character. But rest assured that this research will be done, with direct implications for the wine drinker.

Another unexpected discovery from this kind of mass census is that the winemakers' yeasty friend *Saccharomyces cerevisiae* does not always flourish in vineyard environments. In fact, it rarely occurs naturally on the surfaces of harvested grapes, and is inoculated into them only shortly before crushing, after the wasps have intervened. The strain of yeast desired has to be added to the must, where it can dominate the fermentation process if it is present in adequate quantities. But it is not the only yeast that makes it into the must; wild yeast strains are also present. These uninvited guests may include undesired variants of *S. cerevisiae* that come from the vineyard or that are present in the winery as holdovers from earlier years when different strains were used for fermentation. They may come into contact with the must in a variety of ways, for instance as refugees on winemaking vessels or tools, or by transfer via insects or other animals.

Another category of "wild" yeasts includes species such as those from the genera *Kloeckera, Brettanomyces, Candida,* and *Pichia*. These yeast strains are often important components of wine fermentation, adding to the environmental individuality (terroir) of certain wines. Although such strains and species may do well in the vineyard, they are often less effective during fermentation because they lack high tolerance for alcohol. They also have low tolerance for sulfur dioxide, which is why many winemakers introduce this compound to the early stages of fermentation; they can get rid of the unwelcome yeasts before introducing *Saccharomyces cerevisiae*.

Some winemakers, however, prefer to have the wild strains start the fermentation process. We say "start" because once the fermentation batch attains a 3–5 percent alcohol content the "wild" yeasts will often die, and the more alcohol-hardy *S. cerevisiae* will take over. Balancing the role of *S. cerevisiae* with the activity of wild strains is an important aspect of winemaking. Too little wild influence might fail to impart sufficient terroir to a wine, while too much might introduce contaminant compounds or undesired flavors. This delicate balance is an aspect of wine ecology that winemakers need to control carefully. Finally, we might note that small-scale winemakers sometimes make a point of using native yeasts, which should not be confused with wild yeasts, to produce their wines. Native yeasts might include the yeasts that are hanging around the winery and vineyards, but they might also be domestic strains that are traditional to a particular area.

The enormous microbial heterogeneity on grapes and in vineyards that recent research reveals is causing winemakers to think hard about their farming practices. One study showed that microbial community differences exist between organically farmed grapes and grapes grown the traditional way. Researchers have as yet merely scratched the surface of a huge and wide-ranging field of inquiry; as vine growers and winemakers increasingly use the new microbial community data, they will have an invaluable census of the microorganisms present and active in different kinds of grapes, in different vineyards, and at various stages of the winemaking process. This information promises to revolutionize vine growing and winemaking.

✦ ✦ ✦

We hope we've convinced you that wine is the product of myriad interactions. Those interactions occur at many different levels. First and foremost, wine is the result of an interaction between chemicals and enzymes that leads to its color, nose, taste, and alcohol content. On another level, wine results from the interaction of different microbial species on and inside the grape and in the fermentation vessel. Wine also comes about as the result of the interaction of the parent vine with its environment, particularly with the other organisms living on and near it.

The life processes busily unfolding on the surface of a grape, and later in the grape must, are similar to the interactions going on in a busy industrial region like Elizabeth, New Jersey. Clusters of bacterial cells work together to take in raw materials and make products. The yeast in the must sit on the bottom of the tank and let carbohydrates come to them for processing. They open their bay doors and gather in tons of carbohydrates, which they then disassemble into their constituent parts. Carbon dioxide and alcohol arrive in the mix via the chemical pathways described in Chapter 3. The enzymes the yeast uses to produce the sugars and longer carbohydrates are like tiny machines on a factory floor, and they are continuously fed the raw materials from the grape must. All these interactions produce large amounts of exhaust and waste material, and require an immense amount of energy.

But that's only part of the greater process. There are more than just carbohydrates in the mash. Because the grape skins, seeds, and some stems also get into the must, other little factories spring up in the filmy mats that form from yeast and bacteria. And many larger molecules, such as pigments and tannins, are delivered to those little factories for processing. If a molecule arrives at the wrong enzyme it is refused entrance and will move on to the next potential processing factory. These processes move on apace until the alcohol content of the mix reaches a certain concentration (usually about 15 percent), at which point the yeasts start to shut down. If the alcohol concentration goes too far above this level they will sicken and die. Accordingly, it is at around this point in fermentation that the winemaker moves on to the next stage of production, which involves racking the wine off the solids (pouring the liquid into a barrel and leaving the sediment behind) and, in the case of some reds, subjecting the wine to a secondary fermentation in which bacteria are used to convert astringent malic acid into softer lactic acid (as will be described in Chapter 11). When the wine arrives to rest in the barrel, the process still does not stop. The molecules in the wine interact with molecules emanating from the oak, and even when it is in the bottle in which it will be sold, the wine will continue to change.

Every one of these multifarious interactions, from enticing a bird to eat a grape to aging in the bottle, has combined with the others to make wine both possible to create and such a complex and rewarding product. The unique sum total of these interactions makes each wine you taste individual, different from every other in your experience. And the ultimate interactions are the many that the wine has with us, on its long and complicated journey from our noses to our brains.

7

The American Disease

The Bug That Almost Destroyed the Wine Industry

O n the slopes of Sicily's Mount Etna lies the black, rubbly Calderara Sottana vineyard, locally renowned for the excellence of
the wines it produces from Nerello Mascalese and Nerello Cappuccio
grapes. When the phylloxera insect ravaged the mountainside in the
late nineteenth century, vines in two tiny sections of this vineyard
miraculously survived the infestation. Today these vines, still growing on their own roots, are more than 130 years old, and their grapes
are vinified separately from those of the grafted vines around them.
We had the good fortune to try wines from both sets of vines. They
were obviously close relatives, sharing a mineral-like, earthy quality
with a hint of tar. The regular Calderara Sottana was wonderful, with
dark fruit flavors backed by supple tannins and a lingering finish. But
the Prephylloxera blew us away with the brightness and clarity of
its fruit, and what we can only describe as an extra layer of finesse.

Although the 1860s did not end that way, they began tranquilly for
Jules-Émile Planchon, head of the department of botany at the University of Montpellier, an ancient town in the heart of the southern French
wine country. When he had assumed his chair in 1853, the huge French
wine industry, which in one way or another employed a third of the national workforce, had been in the throes of dealing with a strange fungal
blight. Known as oidium, or powdery mildew, this disease was devastating vineyards across the country. Although viticulturists did not realize
it, the fungus responsible had been introduced from the United States

during an energetic transatlantic interchange of vine cuttings following the Napoleonic wars. Fortunately, the blight yielded to treatment of the vineyards with sulfur compounds, so by dint of heroic efforts it had been eradicated from most parts of France within a dozen years of its appearance. Indeed, the reorganization of the vineyards entailed by the struggle against the disease proved to be something of a blessing in disguise, and the early 1860s turned out to be a boom period for the modernizing wine trade, as it benefited from massive improvements to France's transportation infrastructure.

But the serene conditions were not to last. In July 1866, grapevines mysteriously started dying once more in the vineyards of Saint-Martin-du-Crau, a hamlet near Arles, not far from Montpellier. Green leaves turned red and fell; developing grape bunches withered and dried; root tips began to rot. By the following spring the first affected plants were all dead; within a couple of years, symptoms of the disease were appearing in vineyards throughout the Rhône Valley and the breadth of southern France. As Christy Campbell recounts in his entertaining *The Botanist and the Vintner*, it was clear from the beginning that urgent action was necessary. And in spring 1867 Professor Planchon was appointed to the local Commission to Combat the New Vine Disease. At first, the commission members closely examined vines that had already succumbed to the disease; but even using a microscope, they found no obvious cause. Then Planchon had the idea of pulling up apparently healthy plants growing near the victims. And there was the answer. The roots of these plants were swarming with unfamiliar tiny yellowish insects, all energetically engaged in sucking the sap from their hosts. Planchon immediately concluded that they were the cause of the malady: vampire-like, these insects were sucking the lifeblood out of the plants. Soon he had formally baptized the culprit *Rhizaphis vastatrix:* "vine-devastating root aphid" (aphids belong to the insect order known as "true bugs," or Heteroptera, along with greenflies and plant lice). For technical reasons the creature's official designation eventually became *Daktulosphaira vitifoliae* (finger-sphere of vine leaves), via the also-abandoned appellation *Phylloxera* (dry leaf), the name under which it informally continues to strike chills into the hearts of viticulturists worldwide.

After naming the culprit and tracing its first appearance in the region as far back as 1863, Planchon applied his considerable energies to trying to understand its life cycle. This was no matter of idle scientific curiosity: the best way to eliminate any pest is to find a way to interrupt its development. But although Planchon's meticulous observations taught him a great deal about the bug, he never got the whole story. This is hardly surprising—whereas most insects have only a few stages of development, this one has eighteen. What's more, the phylloxera bug has specialized on the grapevine to such an extent that all those stages are divided into four major life guilds—sexual, leaf, root, and winged—which coincide exactly with the phases of the vine.

As with all insects, phylloxera bugs start out as eggs, which are laid on the underside of burgeoning grape leaves. When they hatch, the emerging nymphs do not begin eating—in fact, they have no mouth or digestive tract to speak of—because their sole purpose in life at this point is to reproduce. The female and male leaf nymphs find each other, have sex, and then die. Before dying, the female lays a single egg in the bark of the vine's trunk. At this point, the sexual guild of the life cycle ends, and the leaf guild begins. Usually laid in the early winter, the egg stays dormant until warm weather returns, at which point it hatches and the nymph seeks out the leaves of the grape plant. The nymph is always female and has the remarkable property for an animal that she can reproduce and lay fertile eggs without having sex. She creates a hospitable environment for herself and her eggs by injecting saliva into the leaf, causing a bulbous gall to form. When these new eggs hatch and the nymphs leave the gall, they either stay on the leaves or make the long trek down to the root of the vine. If they make it to the root, they enter the third guild stage and lay more eggs through virgin birth (technically known as parthenogenesis).

Unlike their sexual guild counterparts, at this stage the nymphs' only goal in life is to eat. As a result, they inflict great damage on the root, especially since one eating strategy involves injecting a secretion that causes the root to soften. This secretion eventually poisons the root, and is one of several reasons for the eventual death of the vine. As the summer proceeds, the nymphs continue to eat and to reproduce through virgin birth for a few more generations. At this stage they can move, though not far, crawling through the soil from one vine to the next. Nevertheless, they can

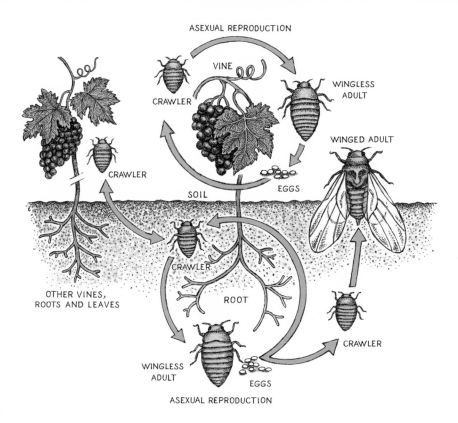

Life cycle of the Phylloxera bug. Redrawn and modified from Murdick M. McLeod and Roger N. Williams, *Grape phylloxera:* Ohio State University Extension fact sheet.

do substantial damage during a single summer and autumn, before winter arrives and the vine becomes dormant.

When the next summer arrives, the insects become active again, and can disseminate in two ways. One strategy is to remain in the same vineyard, in which case the nymphs emerge and lay both male and female eggs on the undersides of the new leaves, starting a new cycle. But alternatively—and this is how the phylloxera bug really gets around—they may enter the fourth guild of the life cycle by developing wings that allow them to fly away and infect new areas. When they arrive, they lay male and female eggs on pristine new vine leaves, and the cycle starts over.

The sheer complexity of this life cycle might make it seem easy to disrupt, but the reverse proved to be true. For the various stages are so bizarre and apparently unconnected that Planchon had extreme difficulty putting his observations together into a clear picture. So when a solution was finally found to the phylloxera infestation, it came from another direction entirely. But in the interim Planchon's technical problems paled in comparison to those he encountered in convincing his colleagues that the insect was the cause of the mysterious malady. Most of the members of the commission on which he served agreed with him that the bug was somehow at fault, but a few influential commissioners felt that its presence on diseased plants merely showed that the plants had already been weakened by some other factor. This was probably climatic or the result of bad viticultural practices, or due to inbreeding caused by propagation via cuttings. Such was the conclusion of eminent entomologists in Paris to whom samples of the bug had been sent, and experts from the dominant wine-growing region of Bordeaux echoed it.

The dispute over the cause of the disease raged for years, even as the southern French wine industry continued its slow-motion collapse, and Planchon worked frantically to find ways to combat the disease in the face of official denial of its cause. During 1870 and 1871, in the midst of the chaos of the Franco-Prussian War and the Paris Commune, French officialdom had greater headaches than phylloxera to contend with, but when these conflicts began to settle it was evident even in the capital that France had a major problem on its hands. A prize was offered to anyone who could find the solution.

By this time the vineyards of the Médoc had also become affected, and the root rot was continuing its exponential spread. Between 1875 and 1889, annual French wine production plunged from 84.5 million hectoliters to a paltry 23.4 million. By the end of the 1870s the ravages of the disease were also evident in vineyards in Spain, Germany, and Italy; and as early as 1873 phylloxera had been detected in previously isolated California vineyards, where the bug had probably already been present for a decade or two. A mere four years later, phylloxera was reported from as far afield as Australia. A major economic disaster was unfolding, not only for the wine industry and the millions of people directly dependent on it, but also for the overall economies of France and Europe and eventually of almost the entire wine-producing world.

By the mid-1870s it was beginning to be widely acknowledged outside Montpellier that the phylloxera bug was indeed the primary problem. One major key to this recognition came from the many ingenious efforts made by French growers to control the disease, which had resisted the sulfur treatment that had driven away the powdery mildew. The most successful such expedient involved flooding affected vineyards during the dormant winter season, a practice introduced by the savvy vine grower Louis Faucon. When his diseased vines rebounded after his riverside vineyard had been inundated for a month by freak flooding in early 1869, Faucon asked Planchon to research the effects of water on the phylloxera bug. The professor showed that little more than three weeks' flooding was enough to drown all the insects and save a vineyard; eventually this simple though labor-intensive approach became widely adopted in France.

Of course, most vineyards are not situated on valley bottoms or carefully constructed terraces that can be flooded and drained at will. Indeed, the majority are on hillsides specifically selected for their good drainage. Nonetheless, although flooding was never going to be a cure-all for the phylloxera blight, Faucon's insight empirically demonstrated the direct connection that Planchon had already made between the parasite and the disease: do away with the insects, and the disease symptoms disappear.

Also crucial to demonstrating this connection was the discovery of where the pest had originated. And again, Planchon was at the forefront of the search. Almost as soon as the phylloxera bug had been reported and identified in France, an Anglo-American entomologist named C. V. Riley began to wonder whether the European sapsucker was the same as the aphidlike creature, now known as *Daktulosphaira vitifoliae,* which the New York entomologist Asa Fitch had found living on the leaves of his home-state grapevines in 1854.

An immediate problem, resulting from incomplete knowledge of the pest's life cycle, was that as far as was known the American insect lived on the vine leaves and didn't cause disease, while the European variety infested vine roots and did. But this issue was partly resolved when Riley determined that different growth stages were involved in the observed root and leaf infestations. Working closely with Planchon, who visited America in 1873, Riley next showed that when an American vine was grafted onto a

Jules-Émile Planchon (*left*) and C. V. Riley

European stock the bugs rapidly descended to the roots and stayed there, killing the vine. Riley's microscopic studies also confirmed that the two insects were identical in both appearance and habits: they were indeed the same bug. It was not lost on Planchon and his colleagues that American vine roots apparently had some feature that the insects preferred to avoid: the bugs confined themselves to the leaves, which might not have been their preferred habitat but from which they could do no long-term harm.

Riley's findings additionally implied—although some resisted this idea, too—that the parasite had been inadvertently introduced into France (at least twice, since the infestations in the Rhône Valley and the Médoc were discovered to have been independent) on vine cuttings imported from America. For although most viticulture in France was proudly based on traditional noble varieties of the Old World vine species *Vitis vinifera,* some curious French vine growers had taken up the cultivation of American vines for both experimental and decorative purposes. These growers included the Bordeaux winegrower Léo Laliman who, at the same congress in 1869 at which Faucon had presented his ideas on flooding, had reported that while he had lost all his European vines that year, his rows of American vines were still flourishing. Imported to test their resistance to powdery mildew, the American vines were evidently also resistant to the new pest.

But there was a problem. Luxuriantly as they might have grown in the Bordelais environment, the American vines produced wines with unfamiliar "foxy" (grape-jelly) flavors that even Laliman had to pronounce atrocious. So, even if the issue of whether the phylloxera bug was the cause or an effect of the disease was not as yet settled, the search was already under way for American varieties better suited for winemaking than those Laliman had planted.

During the mid-1870s, desperate French winegrowers imported hundreds of thousands of vine cuttings from the United States. They did so against official resistance aimed at preserving the traditional noble French varieties. It is hardly surprising that, after the United States had been identified as the source of the blight, the French government fought against the importation of New World vine varieties. After all, how could the problem provide the solution? Still, even as official approval and funding flowed toward such ploys as submersion, pesticide treatments, enhanced vineyard techniques, the introduction of potential predators, and the shifting of vineyard sites to sterile sandy soils, French vines continued to die.

By the early 1880s, even Paris had to yield to the reality that the most successful attempts at phylloxera control were being made by the provincial winegrowers who a decade earlier, at the urging of Planchon and his Montpellier colleagues, had begun to experiment with the officially disfavored American vine varieties. Soon it became evident that the solution to the phylloxera problem would in some way involve American vines, and the question became how.

The most obvious expedient was to identify American vine strains that would produce a better wine than Laliman's had, and there were plenty of candidates from which to choose. Many local grapevine species had been domesticated in the United States during the nineteenth century after efforts to raise European vine varieties had failed miserably there, probably because of phylloxera. What's more, the early introduction of European vines had made it possible for the many native vine species to hybridize with the newcomers, as vines will do, and such mixing events evidently occurred spontaneously on numerous occasions. Promisingly, the offspring of such events tended to combine the qualities of their progenitor lineages. When the American parent was phylloxera-resistant as a result of having co-evolved with the parasite over millions of years (something that Riley, an enthusiastic Darwinian, had noted as early as 1871), the offspring would show at least some of this resistance. At the same time, its European heritage would typically show through in the production of grapes with higher sugar content and a diminution of the foxy flavors that made wine produced from the pure native species unappealing to the Old World palate.

Both hybrid and pure American grapevines of many kinds—it wasn't always clear which was which—were imported into Europe during the 1870s and 1880s. Not all of them successfully accommodated to their new conditions, sometimes failing to root properly, to propagate effectively, or to grow well. The American Vitis *labrusca*, for example, a species from the cloudy, rainy Northeast, failed when planted in the hot, arid vinelands of southern France. Still, in the end half a dozen different robust *américains* became well established both in France and elsewhere east of the Atlantic. Indeed, many European consumers of the inferior wines became accustomed to the foxy flavors that Laliman had deplored—so much so that some French winemakers are even today deeply attached to the American vines and their products, as are many of their clients, including us. Whenever we are in the Dordogne we make a point of visiting a rustic hostelry where the proprietor still surreptitiously grows the American Noah grape, and makes an assertive wine from it that goes particularly well with his braised wild boar.

This does not mean, however, that we aren't happy to return to the much more subtle *vinifera* wines after a brief hybrid dalliance. And, indeed, the américains did not turn out to be universally successful. In the Cahors region, for example, famed since medieval times for its dark, tannic "black wines," the growers who turned to French-American hybrids encountered eventual disaster. Although one apparently accidental cross between the local Auxerrois (Malbec) grape and a variety of the American Vitis *rupestris* produced an acceptable wine, and hybrid grapes kept the local wine industry ticking over in much diminished form for several decades, in the end the wine made from these vines could not compete with the cheap *vinifera* wines that began flooding in from Algeria in the first half of the twentieth century. From 175,000 barrels a year in 1816, wine production in Cahors plummeted to a paltry 650 barrels in 1958. But for the efforts of an unsung hero, José Baudel, who revived the local cooperative, and a handful of small growers who managed to keep the Auxerrois going through thick and thin, the Cahors wine industry would have been finished. Now Cahors stands proud again as a producer of wines that may not be as black as they were before, but that, with the toughness of the Auxerrois now tamed by blending with some Merlot and Tannat, can claim to be among France's most interesting.

The experience of the Cahors growers suggested that almost any viable alternative to the américains would have a good chance of succeeding; and when push came to shove, nearly all straightforward efforts to introduce American vines to France foundered not only on the issue of quality but also in the face of official opposition. The américains were irredeemably tainted by their association with the phylloxera bug as well as by accusations that they were high in methanol—a belief that, unfounded as it may be, still lingers today as your hosts nervously admonish you not to drink too much of that Noah. Government hostility to American vines increased in proportion to their adoption by desperate wine producers, and as the areas of France planted to American vines expanded, official antagonism became implacable. Eventually laws were passed that forbade outright their cultivation anywhere on French soil.

Yet from the beginning of the phylloxera saga an alternative to the wholesale planting of American vines had been available. It depended on the ability of vine cuttings to be grafted onto roots of different stocks. The key factor to a successful graft is that both the scion and the root retain their parental qualities as the plant grows. East Coast North American vine strains had long coexisted with the root-biting phylloxera insect, and many though not all had evolved resistance to its depredations. In its turn, the European species *Vitis vinifera* had been bred over millennia to produce the finest wine grapes in the world. The combination of American roots and European tops thus potentially offered an ideal marriage.

This fact was recognized early on in the phylloxera crisis. Indeed, at the same time Léo Laliman was initially reporting the demise of the European vines in his Médoc vineyards while their American neighbors flourished, he also noted the potential inherent in grafting. Planchon himself was an advocate, and by 1871 his close colleague the Provençal vine grower Gaston Bazille had already joined European tops to American roots. But this exercise proved to be difficult, and progress was slow. In Cahors, for example, it was found that grafting Auxerrois tops to American roots produced vines that were subject to *coulure,* a physiological condition in which the grapes failed to develop after flowering. In the end, it took years of trial and error to find the ideal grafting methods and the best scion-rootstock combinations for different soils and climatic conditions. Eventually, the growers learned that sometimes the best roots might actually themselves be hybrid.

Introducing unfamiliar and labor-intensive new practices to many thousands of vine growers around France took even longer, and despite official encouragement success came unevenly. Still, in the end grafting proved to be the way forward, and today all the noble French grape varieties are grown on roots with American ancestry. Only in a few isolated corners of the world—notably Chile—that managed to escape the introduction of the phylloxera bug are the great European *Vitis vinifera* varieties still grown on a large scale, ungrafted, on their own roots.

For unknown reasons, a few minuscule vineyards in France, Portugal, and Italy also contrived to escape the phylloxera infestation. And all of them have been praised by modern critics for the richness and concentration of their wines compared to grafted local counterparts. Such encomiums lead to the inevitable question: After a century and a half of technical advances in vine growing and winemaking, would the wines of Europe be yet better today if the vines producing them were still growing on their own roots? The reality is that we will never know with certainty, although many twentieth-century connoisseurs were convinced that they would have been. Nonetheless, while for many reasons it would have been much better if the phylloxera epidemic had never happened, the general experience of fruit-tree growers (who have long been enthusiastic grafters) suggests that the grafting process may not make much difference in the quality of the resulting fruit—and hence in the excellence of the wine produced. Still, it's hard to resist a twinge of regretful nostalgia.

✦ ✦ ✦

The phylloxera story didn't end with the defeat of the insect in Europe and elsewhere around the turn of the twentieth century. In an ironic twist, the latest chapter of the saga has unfolded in the United States. For millions of years, California was essentially isolated from the phylloxera-plagued East Coast. So, although wild vines do grow in the western United States, the region was free of the insect when winemaking began in California during the sixteenth century, using the rather undistinguished Mission vine that Franciscan missionaries had imported from Spain. No other varieties were actively planted in California until the 1850s, when *vinifera* cuttings were introduced from both Europe and the eastern states. It was at this point that the phylloxera bug probably first appeared in California, although it was not formally identified there until 1873. The initial affliction spread relatively slowly, perhaps because, unlike its European counterpart, the California insect did not exhibit a winged phase and thus could not readily disperse. After a slow initial response to the disease, the California grape growers agreed that resistant rootstocks were the way forward, and extensive replanting of existing vineyards was done, based on experience in France and elsewhere. So by the time Prohibition came along, the phylloxera bug was no longer a serious issue in the state.

The more recent problem arose when the California wine industry began to boom during the 1960s and 1970s. Suddenly, demand for California wines soared, and growers began not only to bring new land into cultivation but also to search for rootstocks that would be more productive than the purely American Rupestris Saint George variety that most of them were using at the time. Urged by scientists at the University of California, Davis, and excited by its high yields and easy management, growers rushed to plant or replant with a rootstock known as AxR1, a French-American hybrid initially developed in France during the early period of experimentation. The scions it supported produced abundant grapes, and it was easy to graft and grow, but the AxR1 rootstock had been quickly abandoned in France because of low phylloxera resistance. Ominously, it later also succumbed to phylloxera when planted in Sicily, Spain, and South Africa. Nonetheless, Californian scientists and viticulturists either ignored these red flags, or managed to convince themselves that the parasite would not flourish on the AxR1 under West Coast conditions. Driven by visions of enormous productivity, California growers planted huge areas with this rootstock. By the end of the 1970s, up to two-thirds of the vine-growing areas of the Napa and Sonoma valleys were planted with AxR1.

Inevitably, AxR1 vines in a Napa vineyard began to sicken, in 1980. Soon the cause was confirmed to be phylloxera, and the disease raged through the state. In 1989, the experts at Davis issued a rather tardy warning against further plantings of AxR1, but by then it was too late, and by 1992, in the memorable words of the *New York Times* correspondent Frank Prial, "the scene across the Napa Valley was desolate. . . . Piles of dead vines pulled from the soil were being burned. . . . Winemakers watched grimly as their lifeworks went up in flames." The total economic damage was estimated to be about $3 billion, and eventually California wine producers were forced to spend at least half a billion dollars to replant their vineyards with rootstocks of proven resistance.

Fortunately, this herculean effort at phylloxera eradication has so far proven successful; and, for all the trauma it inflicted, the disaster did give wine growers an opportunity to reconsider which varieties were best planted where, and to adjust the compositions of their vineyards accordingly. As a result, the California wine industry has rebounded since the mid-1990s, producing wines that are generally reckoned to be as good as their earlier counterparts.

Perhaps the most important single lesson to be learned from the sad saga of the phylloxera bug and the grapevine is that, if they want to continue making good wine, producers must be constantly on their guard, keeping at least one step ahead of the many organisms that are in competition with wine growers for what the vine has to offer. We can confidently expect that phylloxera will not be the last destructive scourge to infest the vineyards of the world. In addition to all the routine bacterial, fungal, and viral vine diseases, such as powdery mildew, bacterial blight, and leaf scorch, other highly mobile parasites are lurking. A recent bane in California has been the glassy-winged sharpshooter, a lumbering leaf-hopper insect technically known as *Homalodisca vitripennis* that is a vector for Pierce's Disease. This bacterial condition blocks the flow of the xylem that conducts water and dissolved minerals around the plant. An infected vine may die within a couple of years. The sharpshooter is a particularly dangerous vehicle for the bacterium because it moves much faster than even the winged phylloxera bug, and can potentially infect large areas rapidly. Clearly, the price of good wine—or of any at all—is going to be eternal vigilance.

In his excellent *Dying on the Vine,* George Gale, a philosopher of science, makes a point that is of particular relevance in the United States, where in far too many domains we seem to feel that we are not bound by the rules that apply to the rest of the world. He identifies "California exception-alism" as the single most important influence in the unnecessary phyl-loxera debacle of the late twentieth century. Gale quotes one University of California, Davis, expert, writing shortly before the tragedy struck, as claiming that "both the climate and soils of California are natural agencies which tend to reduce the dangers of phylloxera." This insouciant attitude was particularly remarkable given the abundant evidence to the contrary supplied by dreadful experiences that had unfolded a bare half-century earlier. Yes, it can happen here. Or anywhere.

8

The Reign of Terroir

Wine and Place

From the road that winds north through the vineyards from Chagny toward Beaune, you'd hardly notice the low, vine-clad ridge that runs across the near horizon. Known as Mont Rachaz when first recorded as a vineyard in 1252, this is as unremarkable a stretch of landscape as you'll see anywhere in Burgundy. But the wines! The thin layer of limestone intermixed with limy muds that coats the hillside has produced the most spectacular wine-growing terroir anywhere in the world. Years ago, when mere mortals could—very occasionally—afford it, we guiltily splurged on a bottle produced from the Montrachet vineyard. We are still trying vainly to recapture the magic of the moment when we tasted it.

Pity the surface of our planet. It has been constantly battered by the elements since time began. The assault is less dramatic today than it was four billion years ago, during the Late Heavy Bombardment, when asteroids were constantly assailing the newly solidified crust, as the planet mopped up the smaller debris left over in its orbit from the formation of the solar system. But even in today's calmer conditions Earth's surface is under attack daily. Diurnal and seasonal heating and cooling cycles make the continental rocks expand, contract, and crack, while wind and water are constantly eroding them. These merciless forces remove particles from existing rocks and transport them to places of deposition on land or out to sea, where they accumulate. On land the accumulating sediments rapidly become colonized by a vast array of organisms, inaugurating the formation of soils: incredibly complex products of nature that vary hugely from place to place, even over short distances, as the result of intricate interactions among the minerals constituting the rocks, particle sizes, and a plethora of organic influences. And it is with the resulting variety of soils that the concept of terroir begins. Every French speaker instinctively finds a profound meaning in this term, but it becomes curiously elusive when one seeks to translate it into English.

As far as wines are concerned, *terroir* refers most fundamentally to the qualities of any place where grapes are grown. These qualities start with the local bedrock, soils, and drainage, but expand to include such features as slope, exposure, microclimate, altitude, and latitude, along with many other attributes, including the remarkably varying microbial communities we described in Chapter 6. And even after accounting for all those variables we have not explained terroir, because the concept carries resonating echoes of history as well. In addition to physical and biological elements, it also embraces culture and tradition: how local vine-growing and winemaking practices that have evolved over centuries have affected the eventual product of each individual patch of earth. More abstractly, terroir includes the genius loci, that spirit we sense in any magical place.

All in all, then, a wine's terroir is complex and multidimensional, ensuring that every wine in the world is produced under conditions that differ, however subtly or extravagantly, from those in which other wines are grown and made. And there is no question that this makes a significant difference. In the wine world, terroir really *does* matter. In its highest and costliest reaches it is sometimes virtually everything, at least as far as price is concerned. Some of the most expensive land in the world is found not in downtown Tokyo or Manhattan, but in a few little slivers of vine-covered outcroppings overlooking the country road that winds south from Dijon to Chagny, in France's Burgundy region.

There are, of course, those who scoff at terroir and at the reputations stemming from it. They argue that the processes by which grapes are raised and transformed into wine matter more to the final product's quality than the location of the vineyard. And there is some truth to this. In wine, every variable counts; just as one cannot make a great wine without great grapes, bad winemaking can also botch the most perfect of fruit. What's

A view across one of the Montrachet vineyard parcels in Burgundy

more, terroir may be so intensely local that the larger the vineyard becomes, the less likely it is that the average product of the whole will reflect
the qualities of any particular place within it. To benefit from the putative
effects of terroir, it may thus be necessary to strictly limit the size of particular vineyards, or to limit the area within a vineyard where grapes destined to be vinified together may be grown.

This variation is not necessarily bad, and many wine producers benefit mightily from using single-vineyard designations for their wines. It is
certainly a great marketing tool. By common consent, the finest white
wine produced in all of France's Burgundy region—and quite possibly
the world—comes from the vineyard of Le Montrachet, famed at least
since the days of Rabelais. The two communes of Puligny-Montrachet
and Chassagne-Montrachet, a few kilometers south of the city of Beaune,
share this 8-hectare plot. And because of arcane local inheritance laws
it recently boasted eighteen owners, with plots cultivated by twenty-six
local producers. Since some Montrachets now routinely sell in excess of
$3,000 a bottle, few people can boast an intimate familiarity with the prod-
ucts of all these viticulturists. But although there is little doubt that each
year the wines from the various growers will differ somewhat, market

prices suggest that all Montrachets are highly prized by collectors. And with reason: one of us remembers the experience of drinking the 1982 Montrachet from the few rows of vines owned by the Marquis de Laguiche as the most exquisite wine-drinking experience he ever had. Still, even in eight ideally situated hectares there are guaranteed to be differences of soil and exposure, let alone of vinification, and, should we ever win big in the lottery, a horizontal (single-year) tasting of all the wines of Le Montrachet will be high on our list of priorities.

Despite the intensely local nature of terroir, then, it is well worth taking a little excursion into this elusive phenomenon, not least because it is indisputably tied to landscape. Indeed, the landscape in which its grapes are grown is fundamental to any wine, and offers one of the most rewarding aspects of drinking it. There may well be some truth to the claim that one can never fully understand a wine unless one knows the landscape that produced it, and certainly there is nothing to compare with drinking a wine in the place where it was grown and made. Many viticultural regions are breathtakingly beautiful and uplifting simply to be in. It is hard to forget the vistas of the Cape wine lands of South Africa, where neat rows of vines, dotted with whitewashed Cape Dutch farmhouses—the epitome of elegance in simplicity—march up the green foothills toward the blue Cape Fold Mountains. Or the rolling olive hills of Tuscany, which look even today as if they came straight out of a Renaissance painting. Short of visiting such places, what closer experience could you have of them than drinking their wines?

On the other hand, once in a long while, the place where the wine is drunk may prove to be the entire experience. Several years ago, Ian fell for a white wine he tasted in sun-dappled shade by a mountain stream in Nagorno-Karabakh, a tiny enclave in the southern Caucasus. On the face of it, Nagorno-Karabakh is an economically undeveloped backwater with a troubled past, replete with rural scenes straight out of the nineteenth century—peasants nearby were scything spring wheat and loading it onto donkey carts. It hardly seemed a likely place to encounter a wonderful wine. But on a sparkling early summer morning it presented an exquisite environment for vinous enjoyment. In his exhilaration Ian couldn't resist taking a couple of bottles of the heavenly fluid home to New York, and when, after having patiently listened to his lyrical descriptions, his wife pronounced the wine pretty awful, he thought that surely this bottle must have been corked or otherwise spoiled. Alas, the second bottle suggested it hadn't been—and unless somehow it, too, had gone off, the wine really was plonk. The great initial experience had probably been an artifact of the place and the occasion.

It is also possible that the wine was a delicate one, which didn't travel well. (Whether wines can travel is another subject of hot debate among wine consumers.) Ian knew what he thought, though. He had drunk many and varied bottles of French wine on the island of Réunion, half a world away and across the Equator from their vineyards of origin, without ever encountering a bad one. This suggested that transportation by itself, absent some unfortunate circumstance such as overheating—which he knew had not been the case, though it is woefully common among commercially sold wines—had not ruined that Caucasian white. But whatever the case, although the Nagorno-Karabakh story did not have a happy ending, its moral was rather a cheerful one. Place *can* add a positive dimension to the experience of drinking a wine.

✦ ✦ ✦

Vines live a long time. Most winemakers expect their plants to produce for forty or fifty years, and in some places the products of vines more than a century old are particularly prized. So planting inaugurates what will be a very long-term relationship between a vine and the soil that will not only support its roots but also provide it with the nutrients that are essential in determining the quality of the grapes it produces. As matchmaker, the viticulturist must strive to get the marriage right. And although all vines flourish best in well-drained, pest-free, and biologically and chemically well-balanced soils, different vine varieties do better in particular soil types.

For a long time now viticulturists have paid particular attention to the geological maps that indicate the kinds of rocks underlying or adjacent to their vineyards. This is because the soils of vineyards are often derived in large part from the bedrock beneath them, although they are also influenced by other rock particles that may have washed down from higher elevations. Geologists classify rocks into three basic types: igneous, sedimentary, and metamorphic. The igneous rocks of Earth's hard crust have cooled and hardened from molten materials extruded from the fluid mantle below. For the most part, the igneous rocks that form the cores of the continental masses are composed of granites, leavened with volcanic rocks such as basalts and ashfalls. Though they are classified together, the mineral and chemical qualities of igneous rocks may vary greatly. Granites are acidic and rich in hard, erosion-resistant quartz and other light-colored minerals such as feldspar. Typically, soils formed from such rocks are high in rough quartz grains, and thus well drained. In contrast, volcanic rocks such as basalts tend to be more alkaline, and to contain a high proportion of darker-colored iron- and magnesium-rich minerals. The weathering of volcanic rocks generally produces finer clay particles; soils derived from them tend to drain less well.

Sedimentary rocks formed on land consist mainly of compacted particles weathered from preexisting rocks and deposited in other places by the action of gravity, water, or wind. During periods when the landscape is undergoing active uplift erosion rates are high, and sediments accumulate quickly and are often coarse. Such sediments may also be carried into the oceans fringing the continents, and under warm conditions limestones are also often deposited in shallow seas, either by direct precipitation of calcium carbonate out of the supersaturated seawater or by the accumulation of the exoskeletons (basically, tiny shells) of dead microorganisms that had lived in the surface waters. In deeper waters, fine-grained mudstones accumulate.

As the ocean basins open and close through the working of tectonic processes, oceanic sediments of all kinds have routinely been heaved up onto dry land, where today they form the bedrock in many famous wine-growing regions. Oceanic mudstones tend to erode mechanically in much the same way that continental sediments do, but because limestones dissolve in rainwater and are carried away in solution, soils formed on them tend to be shallow and to consist in large part of insoluble impurities from the departed limestone and of various organic residues.

Metamorphic rocks are formed when rocks of the other two kinds are reheated (by the pressure of earth movements or the heat of volcanism) and become recrystallized. It is through this process, for example, that fine-grained mudstones become transformed into harder and more resistant rocks such as shales, slates, and schists, which may also subsequently be weathered to produce soils. Not a few well-known wines, including some Beaujolais from central France, are produced in soils derived from metamorphics.

Water is important in the weathering process, not only because of its mechanical effects and its role in dissolving limestones but also because it encourages the growth of diverse organisms. Plant roots penetrate cracks in the rock and enlarge them as they grow, and lichens can begin chemically transforming the decomposing rock at an early stage of soil formation. But the influence of water extends yet farther, since this transport medium can sort rock particles by size. Rapidly moving water can carry along large pieces of detritus, even boulders, until the energy propelling it is lost and the particles are deposited in places where they can accumulate. Slow-moving water, on the other hand, will carry along and eventually deposit only the finest material. Although time is also an important factor, if other things are equal, the kind of soil formed when the sediments are exposed at the surface will depend heavily on the particulate composition of the underlying sediments. Coarse river gravels deposited under conditions of moderately high energy will provide a very different medium for vine growth than will fine clays left behind on a dry lake bed.

But geology is far from the whole story in soil formation. Local climate, in terms of both temperature and precipitation patterns, also exerts a huge influence on soil formation. So do independent considerations such as slope, exposure, and even position on a slope. Generally, soils at the top of a slope will be better drained than those farther down, and this too will show up in soil composition, with higher organic accumulations typical downslope. Finally, time also has its effects: the longer a soil has been forming, the deeper it will tend to be and the more its profile will reflect circumstances prevailing at varying times in the past. All in all, soils are dynamic, always in a state of flux.

Because growing vines always has the potential to degrade soils, vine growers in long-established viticultural areas have learned how to minimize these effects—although they seek as much to discourage excessive vigor in the growing plants as to promote it. If there is too much nitrogen or water in the soil during the spring, for example, the vines will produce too much foliage, and the fruit, developing in the shade, will ripen too slowly and yield tart, underdeveloped flavors in the wine. Faced by excessive foliage the winegrower can, of course, try to correct the situation by leaf removal, but this is a labor-intensive process that may have undesired consequences, and it is much better to eliminate the problem at its source if possible. Often this means avoiding particular soils and exposures, as is still done in Burgundy, where some of the most expensive vineyard land in the world is interspersed with otherwise inexplicable tracts of forest regrowth.

In Burgundy, hundreds of generations of viticulturists have laboriously discovered where to grow and not to grow vines, over many centuries of trial and error. This is one reason why the "Bourgogne" (Burgundy) appellation on a bottle of wine may indicate something significant about its contents. But the appellation system in France also pertains to other factors, including the grape variety or varieties allowed and the permitted fruit yield. Variables like these may be just as important as indicators of wine quality as the particular spot on which the grapes were grown. What's more, even where particular vines have been grown for the longest period of time, optimization is not necessarily assured. For proof of this, look at the dozens of Cabernets and Merlots now on offer in the United States from areas of southern France where, until recently, wines from different grapes were produced for two thousand years. A few decades ago, the vignerons of this region would have been appalled at the very idea of growing anything other than their traditional cultivars, and the recent shift toward grapes whose names are well known to American consumers must presumably have been influenced at least in part by marketing considerations.

Whether terroir was a factor in the change, and whether the introduction of the new varieties was a good idea, time will probably tell; but meanwhile, how can a vine grower, faced with planting a new vineyard or even with replanting an old one, predict which soils will be good for which grapes, and whether interventions such as irrigation or fertilization of a particular location will improve the prospects for producing excellent wine? Sadly, scientists have provided less helpful information than might have been expected. Most of what is known is local, carried around in the heads of experienced vine growers who may or may not pass it along to the next generation (the same can be said, by the way, for other important inputs into winemaking, such as the choice of wood for barrels). Still, there is no shortage of consultants.

The most important confounding factor in the scientific study of terroir has been the difficulty of ruling out the relative influences of multiple variables. One research project in South Australia did find significant differences in yield, acidity, and color between Shiraz grapes produced in irrigated and non-irrigated areas. The quantities of wine produced by the dryland vines were considerably lower, but their acidity and concentration were higher, and their coloration was more intense. Small wonder that these were the wines preferred by a tasting panel. Such findings reinforce the common wisdom that reduction of yield by stressing vines (which in this case had to work harder to obtain the water they needed) can produce a superior wine. But nonetheless the distributions of soil types proved to be so complex over the regions involved that variations in soil depth and composition, important components of terroir, could not be ruled out as significant factors in determining wine quality.

One group of German scientists tried to address this variable by growing Muller-Thurgau and Silvaner vines at a single location, in tubs filled with seven different kinds of soil. They found no significant differences among the wines they eventually made from these vines, but it would be premature to conclude that under normal circumstances soil type and quality make no important contribution to wine quality. The containers used in the experiment were unavoidably limited in size and provided unnatural environments, and these artificial factors, in and of themselves, were probably important determinants of the wines grown in them.

Here, then, we have a clear example of the observer effect, in which by scrutinizing something the observer changes it. In addition, experimental interference of the kind scientists are accustomed to doing in the laboratory only exacerbates the real-world problem. In any complex system, trying to control for one variable inevitably affects a host of others. Even advanced instrumentation that measures numerous different qualities in a soil sample, or of a microclimate, cannot capture terroir; it still yields only some of the attributes of a particular location. What it does not do is define what makes the location special as a place for growing wine. Complicating the problem is the subjectivity of the study: people's idea of what makes a wine good, great, or appalling can vary. Let's look at a couple of examples to see how mystifying the idea of terroir can be in practice.

◆ ◆ ◆

Bordeaux is one of France's great urban centers, lying on the Garonne River just above the spot at which it meets the Dordogne River to form the Gironde Estuary and flow into the Atlantic. The city has lent its name to a large vine-growing region (the Bordelais) that both surrounds it and extends across the two rivers. In the east, beyond the Dordogne, a limestone escarpment crops out, while the city itself, and most of the vineyards surrounding it, lie on a thick pile of gently undulating and hugely complex river-lain deposits that change in precise aspect at practically every step. The vineyards of the Médoc, the famous winemaking region along the left (west) bank of the Garonne and Gironde, lie mostly on large mounds of river gravels that incorporate lenses of finer clays, silts, and sands. The coarse gravels, mainly originating in outwash from an early deglaciation, contribute a well-drained structure to the soils in which the vines grow, while the finer sediments both trap moisture and contain essential mineral nutrients. At the surface there is little except rather sterile gravels, which explains why deep-rooted vines grow best here—many renowned vineyards lie on otherwise "poor" agricultural land. Below the surface, however, conditions are different. The vines flourish as their roots find the clayey lenses and send rootlets into them, even as they continue striving single-mindedly down through the more sterile gravels toward the water table below (the stress factor again).

As might be expected from the geological complexity and variety of the Médoc sediments, not every stretch of soil is equally good for vine growing. And over the centuries it has escaped neither vignerons nor wine buyers that some vineyards produce better wines than others do. As early as 1855, the market for wines from the Médoc had become so evolved that a classification of its finest vineyards was created for the Exposition Universelle in Paris, based on the prices their wines had historically fetched at auction. Four vineyards (one of them actually to the south, in the Graves region, but with a similar geology) were designated Premiers Crus (First Growths). These were the four that had for long been most famous and whose red wines sold at the highest prices; a fifth Médoc vineyard has since been

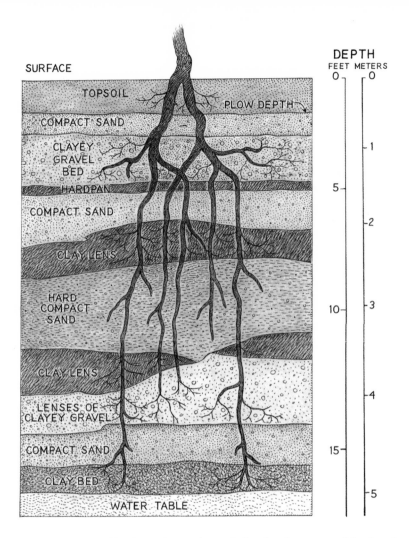

Generalized soil profile of a Bordeaux vineyard, showing the types and depths of sediment through which vine roots penetrate there. Redrawn and modified from a diagram in Seguin, *Influence des facteurs naturels sur les caractères des vins*.

promoted. Eleven vineyards were designated Deuxièmes Crus, "Second Growths," and a further dozen or so were placed in each of the Troisième, Quatrième, and Cinquième Cru (Third, Fourth, and Fifth Growth) categories. In this case, price was the explicit criterion for quality, but implicitly the differences in ranking (or whether a vineyard was ranked at all) represented differences in terroir. Of course, the winemakers were different, too, and much else has also changed since 1855. But despite occasionally vociferous complaints, the classification has survived remarkably unchanged for a century and a half, and the Premiers Crus still fetch by far the highest prices, trailed by the "Super Seconds" (a modern coinage) at the top of their category. It is possible that what we are observing is actually success breeding success: the vineyard owners whose wines fetched the highest prices were best able to invest in the upkeep of their land and in the finest winemaking techniques and equipment. But the durability of the classification, over a long period during which winemaking processes have become significantly refined, and ownership of the vineyards has repeatedly changed hands, suggests that place—terroir—is one controlling factor.

So what is special about the highly ranked vineyards of the Médoc? Given that their soil profiles vary so intensely, on such a microscale, probably few if any of them could be said to possess only a single terroir. But some generalizations can be made. Almost all of the wines that made it into the 1855 classification come from three communes—Margaux, Saint-Julien, and Pauillac—that are strung in a line along the left bank of the Gironde Estuary, toward the north of the large Médoc winegrowing region. All four of today's Premiers Crus from the Médoc are from Margaux and Pauillac. Looking only at the gravelly surfaces of the vineyards that are typical throughout the Médoc, a human observer wouldn't notice much difference among them. But the deep-rooted vines clearly do. The three most highly reputed communes of the Médoc are all on deep, well-drained gravel banks that are interspersed with crucial clay and silt lenses. Toward the north into Saint-Estèphe, also an esteemed wine producer, the clays begin to increase as a proportion of the sediments, impeding drainage. Vines hate getting their feet wet, and to the north of Saint-Estèphe, where the soils begin to get really heavy, the vineyards start to peter out.

To the south of the three key communes, the gravels increase. Yet many claim that the vines of southern Margaux, with its thinnish soils atop relatively coarse gravels, produce on balance the finest red wines of all, especially in rainy years when drainage is most important. Throughout the Médoc the vines have to work hard to find the combination of moisture and nutrients they need, but in Margaux, depending on conditions, they have to work perhaps hardest of all. Yet the flagship estate of Margaux, the revered Château Margaux itself, seems to be a bit of an anomaly. According to the geologist James Wilson, the most prized tract of the Château Margaux vineyard, a feature called the Cap de Haut, lies essentially on a freshwater limestone bedrock, which means that the roots of the vines have to penetrate through cracks in the rock and nourish themselves from impurities they encounter. This situation recalls the areas of Saint-Émilion and Pomerol to the east of the Dordogne River (with prices to rival or even exceed the Médoc's), where visitors can descend into wine cellars from which building limestone has been quarried and peer up at the roots of vines penetrating down from above. Lower-priced (though still highly prized) white wines are grown in a section of the Château Margaux vineyard that lies on exposed marine chalk.

Wines from the Margaux appellation are often characterized as suppler and more "feminine" than those that come from around Pauillac to the north. Pauillac reds are famed for their "masculinity" and tougher, more tannic structures, though they display no less finesse. Here there is a bit more topography and more variety in substrate. The Premier Cru Château Latour lies on the "southern lobe" of the Pauillac gravels, separated from most of the classified growths and close to the waters of the Gironde. The gravel terrace on which it lies is exceptionally thick. It contains some very large stones as well as clayey and silty lenses, and also overlies some marly sands that are rich in oyster shells. The Premier Cru Châteaux Lafite-Rothschild and Mouton-Rothschild lie some distance away on the northern lobe, farther from the water but once again on thick gravels with good drainage. So the signal is fairly consistent: to get the best fruit in this fabled vine-growing area it is necessary to grow the vines in deep, well-drained soils, with a nonetheless accessible water table and finer-grained lenses to harbor moisture and nutrients for the growing plant. Roots love stable conditions, and when all these elements are present the vines can send their roots far down into the substrate, sampling the greatest variety of soil components possible while insulating themselves from the more changeable soil environments near the surface.

Bearing all this in mind, Gérard Seguin, a researcher at the University of Bordeaux, suggested in the 1960s that the ideal situation for a vine in the Médoc is above an ancient drainage channel that had initially been carved by river flow but had then been filled up by flood deposits as the main channel moved away. In such a situation the subsoil will be unusually dry, forcing the vines to send their roots down many meters, passing from time to time through silty lenses and sending out more rootlets. The corollary of this idea is that the Premier Cru vineyards are the ones closest to those deep drainage channels, while the lesser crus are progressively farther away. As far as the vines are concerned, what is under the soil is often at least as important as the soil itself.

Although the nature of the surface seems not to be a critical factor in Bordeaux, this cannot be said of everywhere. An early rival to the "clarets" of the Médoc (a corruption of *clairette*, the name given to the dark rosé wines that Bordeaux wine merchants mainly exported to England until the eighteenth century) was the famous and very differently styled black wine of Cahors. Dark and brooding, made from the Malbec grape (locally known as Auxerrois) that is today only a minor component of some Bordeaux blends, this wine was grown far inland along the valley of the Lot River, a major tributary of the Garonne. The Lot flows through a magnificently eroded limestone landscape and, far away from the moderating influence of the ocean, Cahors enjoys a more extreme climate than the Bordelais does. As a result, anything that will help buffer the soil from typically wide swings in temperature and humidity is welcome to wine producers in the region. Excellent wines are grown on the alluvial gravels down in the valley of the Lot, but by general consent the best Cahors wines come from high on the valley sides, or from the heavily weathered plateau above where the iron-rich limestone soils are well drained and flattish limestone pebbles abound on the surface. These pebbles both preserve and distribute moisture, and by protecting the soil below from extreme excursions in humidity and temperature, they enhance the stable conditions that the vine roots prefer. Light in color, they also reflect sunlight and warmth back up into the foliage of the plant and onto the leaf-shaded grapes, promoting the ripening process. A difficult earth surface that would be a nightmare for any other kind of agriculturist is a boon to a Cahors vigneron!

✦ ✦ ✦

Perhaps the most famous vine-growing region in the New World is the Napa Valley of northern California. Lying at the edge of a vast continent that has for millions of years been pushing with unimaginable force against the adjacent tectonic plate, the region has a fantastically complex geology. The Napa Valley proper (the winemaking region is actually larger) is a flattish area, up to 5 kilometers wide with significant topography, extending for around 50 kilometers between Carneros and Calistoga. To its south and west lies the Mayacamas mountain chain; the Vaca Mountains define it to the north and east.

The Vacas are formed primarily from rocks belonging to the Napa Volcanics, as are the bumps in the valley bottom. Similar rocks are also found locally along the other side of the valley, but most of the Mayacamas range is made up of rocks belonging to what is called the Great Valley Sequence, a contorted pile of mainly marine sandstones and shales. The upshot is that the many kinds of rock contributing to the walls of the Napa Valley provide a diverse set of sources for the sediments in the valley bottom, augmented by others brought in from far away by the Napa River.

In *The Winemaker's Dance,* the geologists Jonathan Swinchatt and David Howell propose that all the sediments in the valley bottom were actually deposited quite recently (in geological terms), their predecessors having literally been washed out by the Napa River at a time when huge quantities of Earth's water had become sequestered in the greatly expanded ice caps that formed during the last glacial period. Peak ice expansion occurred around eighteen thousand years ago, and this sequestration of water caused sea levels to drop to some 100 meters below today's shoreline, leaving San Francisco Bay high and dry. In this process, new energy was imparted to the Napa River, allowing it to flush out huge quantities of sediment from the valley on its way to joining the Sacramento River and the sea. Swinchatt and Howell suggest that, as a result, today's superfi-

A view across the rolling Napa Valley vineyards

cial valley sediments were probably deposited no more than ten to fifteen thousand years ago, so the soils formed on them have not had time to develop mature profiles.

With immature soils on its floor, and thin ones clinging to its steep and heavily eroded sides, the Napa Valley would hardly seem the most propitious place in which to produce some of the world's greatest wines. Yet anyone who has tried one of Napa's better products well knows that this is indeed the case. Still, even with intense interest from investors in neighboring San Francisco, vines do not cover the entire valley. Terroir clearly intrudes again, and sedimentary origin is obviously an important consideration.

To address this issue, Swinchatt and Howell proposed a broad classification of soils in the Napa Valley into three kinds: residual, alluvial, and fluvial. Residual soils developed on sediments that are still sticking to the sides of the surrounding hills. Alluvials were formed as sediment-laden streams descended the valley sides, reached its floor, lost energy, and shed their sediment loads to create alluvial fans. The fluvial sediments were directly deposited by the Napa River itself.

According to Swinchatt and Howell, residual soils are poorly evolved, extremely well drained, and sparse in nutrients. These features create a stressful environment for vines. Typically, wines made in the mountain vineyards offering these conditions are tightly structured and highly concentrated: tough when young but capable of elegance with age.

The alluvial sediments along the valley edges sometimes extend quite a way toward the river and boast gravels, sands, silts, and clays in various proportions. Some of the most renowned vineyards at the bottom of the Napa Valley are situated on the "benches" formed atop these fans, but not all alluvial deposits are equal, and too much fine sediment or inadequate soil depth will prevent the vines from producing the best grapes. Some of the finest alluvial vineyards are situated toward the valley side, where the sediments are on balance coarser. Good drainage thus appears once more to be the key, so long as there are also adequate nutrients. But, again, differences on a microscale appear crucial. This is because some alluvial contexts produce spectacular wines that tend to vary among themselves in character rather more than mountain ones do, while others produce plonk.

Fluvial sediments are the most difficult for the vine grower, perversely because they are typically rich in nutrients and encourage excessively vigorous growth in the leafy parts of the vines. It is difficult to manage vines that overgrow, and Swinchatt and Howell note that no highly regarded wine is produced in Napa from fluvial-grown vines alone. They also observe that in general, vines in this setting produce a rather herbaceous-tasting product.

It appears, then, that little as we can say scientifically about why one site may be outstanding for viticulture while a nearby site is not, soils are obviously a critical factor in producing excellent wines. A good substrate is a necessary condition for growing grapes with the chemical composition that a first-class wine needs. But clearly it is not a *sufficient* condition for producing the very best—or even for growing exceptional grapes. Just as there is many a slip between planting vines, harvesting grapes, and producing wines, there is a lot more to terroir than merely soil. High among those additional factors is climate.

✦ ✦ ✦

Terroir is the essence of place, and one of the most important aspects of any place is its climate. This, though, can be tricky. Broadly defined, climate is the weather at a specified place on Earth's surface, averaged out over the year—or over decades—and expressed in terms of temperature, atmospheric pressure, precipitation, cloudiness, wind speed, and a host of other variables. Each variable is influenced by many factors, including elevation, latitude, topography, and proximity to water. But just as the average human is said not to exist, neither does the average day, nor indeed the average year.

Vines can be grown in an amazingly wide range of places: we have enjoyed wines produced right on the Equator in Kenya's Rift Valley, while Alaska currently has at least four wineries. But given the grapevine's geographical origins, it is not surprising that vines do best in the temperate and Mediterranean climatic zones between about 30° and 50° latitude. Temperature, closely related to sunshine, is particularly important in regulating the basic physiological processes on which vines depend, such as respiration and transpiration, since many of these are largely inactive below about 10° C. In high temperatures ripening tends to occur rapidly, burning the plant or building up the sugars too quickly at the expense of other important compounds; if temperatures are too low the sugars may not develop fully, and the acid components of the juice will predominate. Frosts can damage the plants, especially early in the season when they are putting out new shoots and buds, while winter deep freezes can kill vines outright. Rainfall is another critical factor, especially during the growing season when too much rain may encourage the growth of mildew, or dilute the grape juice if it falls too close to harvest. In places with less than about 70 centimeters of rainfall a year vines may need irrigation, although surface watering (like fertilization) may discourage the roots from going deep. Winds, too, are important, sometimes cooling the vines and in other places warming them.

Fortunately, there are lots of places that are congenial to vines' basic needs. But there is no doubt that better wines are made in some places than in others, and that climate plays a big part in this. One way of coping with differences in regional climate is to grow varieties with different climatic preferences, and it is no accident that grapes such as Silvaner and Chardonnay are favored by growers in Germany and northern France, whereas in southern Spain they grow the Chipiona and the Chiclana—both of which did particularly well in the hot and extremely dry growing season of 2012, though yields were low. Yet wonderful Cabernet Sauvignons are made in places as diverse as Bordeaux and Napa, and great Pinot Noirs come from such disparate areas as Oregon and Burgundy.

More important, even within these limited areas both microclimates and the quality of the product may differ enormously. Topography is always a vital factor, and because of that, compromises have to be made. If a vine grower decides to site a vineyard high on a hillside to obtain better drainage, for example, the angle of each vine to the sun and the amount of sunlight each receives will necessarily be different depending on the vine's position on the slope. If the plot curves tightly across the hillside, each row of vines will have its own individual exposure. How high it is up the hillside, the slope of the hill, and how fast air settles down it will all have an influence on the microclimate enjoyed by an individual plant, even leaving out considerations of the nature of the soil itself. And this brings us back, as always, to that mysterious factor of terroir.

On a large scale it is usually easy to say what generally happens climatically in a great wine-growing region. The Bordelais may be at roughly the latitude of Nova Scotia, but it is topographically low lying and is directly adjacent to the Atlantic Ocean, which provides the moderating influence of the warm Gulf Stream. The prevailing winds come in from the west, picking up moisture over the ocean and providing rain year-round as well as fogs that moderate the direct effects of the sun. Forests along the shoreline protect the vine-growing areas from salt-laden low-level winds. In the Médoc, winters are usually cool without being excessively cold, and although the summers can be warm, lack of sunshine may be a problem because of cloud cover: the greater vintages in Bordeaux are normally reckoned to come in the hotter years. Although the Merlot grape dominates in the east of the Bordelais, and the Cabernet Sauvignon in the west, the local winemaking tradition is to insure against climatic vagaries by blending different grape varieties, each with its own ripening characteristics. The generally easterly facing vineyards of the Médoc cluster toward the Gironde while remaining high enough to have adequate drainage. But given the rather flat topography of the area, solar exposure seems to be a relatively unimportant factor.

The Napa Valley, which lies 7 degrees of latitude to the south of Bordeaux, offers a complete climatological contrast. Hemmed in between mountain ranges that separate it from the Pacific Ocean to the west and the semidesert Central Valley to the east, Napa experiences cool and wet winters, along with hot summers that are somewhat moderated by dense fogs. These fogs result from the contact of warm, moist Pacific air with the cold Humboldt Current right offshore, and they are drawn inland over the Napa Valley by air currents rising from the hot floor of the Central Valley. Within this broad context Napa's irregular topography produces a host of microclimates, with dizzily varying exposures, slopes, and altitudes. Compared to the Médoc, these microclimates are relatively stable from year to year, ensuring a reasonably consistent product and creating a situation in which vine growers can concentrate on the varieties most appropriate to each parcel of land.

Yet Napa is significantly warmer than Bordeaux, which might be a cause for asking whether the Cabernets and Merlots for which Napa is famous (both varieties that were developed in cooler climes) are the optimal grapes for the valley. Perhaps varietals more typical of southern France, or even Spain or Sicily, might be more appropriate? A good point, but it can be argued both ways. Yes, there is a lot of Cabernet produced in the Napa Valley that probably should not be. Many Napa Cabernets come across as excessively fruit-forward and lacking in the tannic structure that gives the best Bordeaux their harmoniousness and elegance. But Napa is home to some fine very Cabernets—many of them grown on the cooler upper slopes of the fringing mountains, and on volcanic soils that are a far cry in structure, origin, and exposure from those of the Médoc. Indeed, an informal comparison we conducted in 2013 of several older vintages of the Dunn Howell Mountain Cabernet, grown at high elevation near Calistoga at the northern end of the Napa Valley, and Cabernet-dominated blends of the same vintages from the Super Second Chateau Lynch-Bages in Pauillac showed a remarkable degree of convergence between the two wines. This was not quite what we or our host Mike Dirzulaitis had expected, and although it was close, in most years the California wine got the nod.

✦ ✦ ✦

So what are we to make of terroir? Well, all winemakers and wine lovers agree that there are better and worse places to grow wine grapes. But the differences between different vineyards can be traced to a practically endless array of factors that range from the physical medium in which the vines are grown to the latitude, altitude, and exposure of the particular locality. In addition, different types of vines do better under different growing conditions, again in terms of both soils and microclimates. The size of the plot will also affect the assessment of the wine it produces—in extreme cases, a specific terroir may not be a much bigger than a large tablecloth. But that's just the start. The ultimate arbiter of terroir is the excellence of the wine, which is also deeply influenced not just by where the vines are planted but by the ways in which they are pruned, trained, irrigated—or not—and even by how far apart they are spaced and by the local microbiome. And all this is merely before the grapes (oops, how long did you wait before picking them?) get to the winery, where they may be crushed, vinified, and matured according to any number of different protocols. So much happens between planting a vine, and drinking its product, that separating out the influence of any one of the factors involved along the line is next to impossible.

Still, only zealots would deny that terroir has to mean something. And if we look just at the *terre* (earth) part of it, maybe we are missing the point. Terroir has a larger meaning, which embraces all the physical, biological, and cultural aspects of winemaking. Above all, it is the vines that know where they like to be, and where they can do the best job. Of course, they are mute about why they like to be where they are, or why they don't. But they do express their views in what they produce, and for a long time now, people have been listening to them. So let's be grateful to the generations of winemakers who have listened and have identified the best places in their corner of the world to grow their grapes and the most effective ways in which to vinify them. And let's be grateful too, to those who have worked hard in our own times to identify prime sites and to produce the finest wines they can from them, using the best techniques they can devise.

In the end, what is most fascinating about wine is its sheer variety, something that derives from many sources, including any component of terroir you might mention. As a result, the pursuit of fine wine is a dynamic preoccupation that is always changing, and has many ways forward. Romans today no longer spend their birthrights trying to obtain wines grown on the slopes of Mount Falernus; this now-obscure spot is just another of the many places in Campania where they have forgotten the Aminean vine and grow Aglianico instead. Maybe in a thousand years' time Le Montrachet will be just another vineyard where Chardonnay (or something else) is grown; perhaps it will have sprouted a shopping center, and people will prize the Chardonnays from Mount Kilimanjaro above all others; or maybe it will still be the greatest vineyard on earth. Only time will tell. But meanwhile, we are still hoping to win that lottery.

9

Wine and the Senses

W hat wine appeals more comprehensively to the senses than Champagne? It sparkles in the glass; it hisses slightly in your ear; its fragrance varies all the way from nutty to ripe pears to freshly toasted brioche; it hits your palate with a fine effervescence; and a great Champagne teases your tongue with a cascade of sensations before fading away with lingering slowness. Every sense that a wine can reach, Champagne will. Many fine sparkling wines are produced worldwide, not least in Italy and California; but once in a while there is nothing like returning to a fine bottle of Champagne.

Galileo Galilei is best known for his novel way of looking at Earth's place in the solar system and his consequent problems with the Vatican. But long before all the fuss blew up over Galileo's cosmology, he had produced a remarkable work called *Il sagiatorre* (The Assayer). Published in 1623, it ranged broadly across the sciences, with a focus on vision. And the science historians Marco Piccolino and Nicholas J. Wade have recently pointed out how innovative Galileo's philosophy of perception was. Among other things, Piccolino and Wade quote Galileo as claiming that "we should realize quite clearly that without life there would be no brightness and no color. Before life came, especially higher forms of life, all was invisible and silent although the sun shone and the mountains toppled." Galileo was saying that while the physical attributes of the planet are present, they are perceptually nonexistent until they have been interpreted by our senses. This theory applies to wine as much as to anything else, and Galileo, who described wine as "sunlight, held together by water," did not forget that fact. As he put it in *Il sagiatorre,* "A wine's good taste does not belong to the objective determinations of the wine and hence of an object, even of an object considered as appearance, but belongs to the special character of the sense in the subject who is enjoying this taste."

What Galileo was perceptively telling us was that, to describe what a wine tastes, feels, looks, sounds, and smells like, we need to understand how the senses work. Anyone who has ever attended a wine event knows the five S's of wine tasting: See, Swirl, Sniff, Sip, Savor. The five S's allow us to hit directly three of our five senses—sight, smell, and taste. This leaves us with two senses that we rarely associate with wine—hearing and touching. But ignoring them is a mistake. There are few things more satisfying than the classic Pop! of a Champagne bottle, however déclassé purists may consider it (they prefer an unostentatious hiss). More important, what a person has heard *about* a wine usually influences his or her perception of it. In fact, the multimillion-dollar wine advertising industry depends on this aspect of wine appreciation. As for that fifth sense, touch is also critically important in how we perceive wine—not through our fingers but through touch sensors in our mouths and throats. If we couldn't *feel* the wine in our mouths, our experience of it would be incomplete.

❖ ❖ ❖

Let's start with vision. Color is critical to the appreciation of any wine, and the presence of pigments in grape skins may derive from the vines having evolved traits to attract birds (which have exquisite color vision). Eyes have evolved more than twenty different times among living creatures on this planet, but it is a good bet that birds' eyes and our own have a single common origin, and they certainly have many functional similarities. This being so, it is reasonable to suppose that in some sense we too might be predisposed by our biology to be attracted to the various colors of the grape. Humans appear to prefer reds over blues, greens, and yellows, and how we perceive red is important in the formation of our preferences in wine.

Light has had a complicated history of study. It was thought by some to be a particle and others to be a wave; in fact, the best way to describe light is both as a wave *and* as a particle. But it is the wave nature of light that allows our eyes to detect specific colors. Things appear to have different colors to us because our eyes and brains can detect very small differences in reflected light waves over a very small part of the wavelength spectrum. Visible light ranges from a wavelength of 0.4 micrometers (0.4 millionths of a meter) at the violet end of the spectrum to 0.7 micrometers at the red end. White light is a mixture of all of these wavelengths. Different colors in between occupy specific wavelengths within the spectrum of visible light.

Our perception that wine and other objects have color comes from the wavelengths of light reflected from them or passed through them. Ambient white light is made up of all the colors of the spectrum—red, yellow, green, blue, indigo, and violet. When we see something as white, we are actually seeing all of the colors of the spectrum fused together. And how an object appears to us is determined by which portion of this rainbow of colors it absorbs, or reflects. For instance, when white light hits it, a red grape absorbs all the colors of the rainbow except the light at the red end of the spectrum. That is reflected, so red is what we see. Similarly, a so-called white grape (actually, a green or slightly yellow grape) absorbs all the colors of the spectrum except the light in the green and yellow range.

The reflected wavelengths impinge upon sensitive cells in the retinas at the back of our eyes. And the story from there on out is largely a molecular one. The retina is like a cornfield full of long, thin cells called rods and cones. These are connected to nerve cells that are "wired" to a region at the back of the brain called the primary optic area. The rods and cones lie in close proximity to one another, but are structured differently and support different populations of proteins, which in their most relaxed state are simple linear molecules that look like beads on a string.

The workhorse of visual sensing is the category of proteins known as opsins. The opsins anchor themselves to the cell by winding through the cell membrane seven times. This interweaving leaves parts of the beads on the protein string exposed on the outside of the cell, while others are on the inside. When hit by light of specific wavelengths, a specialized part of the outside beads causes the protein to flip, from a form called *cis* to a form called *trans*. These flips are incredibly precise, and correspond to the exact wavelength of the light that has hit the retina. The jolt causes a chain reaction within the cell, and this is transmitted as an electrical potential to the nervous system and on to the brain.

The membranes of our rod cells contain rhodopsins, a specific type of opsin. When monochromatic light (light of a very narrow wavelength) hits the retina, the rod cells are stimulated. At night, all light is transmitted to the eye as monochromatic, so rhodopsin is an important component of night vision. In contrast, our cone cells have a choice between four different kinds of opsins, giving us four different kinds of cone cell. These four opsins are conveniently named long-, medium-, and short-wave sensitive, or LWS, MWS, and SWS1 and SWS2. Each of the four types of cone cell in the retina is like a switch that triggers a specific part of the brain to recognize that a particular wavelength of light has hit the eye. The LWS opsin detects light in the red range, the MWS detects light in the green range, and the two SWS opsins detect blue and violet.

Because of the versatility of these opsins in detecting light of different wavelengths, most human eyes are sensitive to subtle changes in color. But they can see them only if the opsin proteins that detect the different wavelengths of light hitting them are working properly; and most people reading this book will probably know someone who is red-green color-blind. (One out of every eight males of European descent has this condition.) These individuals cannot discern between the red and green colors hitting their retina, and hence cannot tell the difference in color between a glass of red wine and a glass of crème de menthe (without smelling it).

Individuals with only two of the four cone cell opsins will have dichromatic color vision. Only light that has wavelengths that excite the two kinds of opsins will be visible. In fact, most humans are considered trichromatic, even though they have all four opsins. This is because one of the SWS opsins is blocked by absorption, and hence is rendered nonfunctional. In the past decade, vision specialists have started to find individuals (all female) who are truly tetrachromatic, with four fully functional kinds of cone cells. These individuals see arrays of colors that are orders of magnitude more bountiful than the shades and hues in the 136 Crayola crayon box. Researchers have estimated that the addition of the functional fourth kind of cone cell allows these individuals to discern from a hundred to ten thousand times more colors, hues, and shades than trichromats can.

Wine appears to be red when it is full of anthocyanins, the family of chemicals which absorbs a specific wavelength of white light. Over 250 different kinds of anthocyanins have been found in plants, and they act as light sponges. They sponge up light most efficiently at wavelengths of about 520 micrometers. This means that all the green and yellow light is absorbed, leaving light waves of above 620 micrometers to be reflected to our eyes. The flip side of absorbance is transmittance: if, for example, the light transmitted through a glass of wine had a wavelength in the range of 650 to 700 micrometers, the wine would be very red.

If we were all tetrachromatic, we would easily detect extremely subtle differences in color—and would have developed a complex vocabulary to suit. But most of us are not, and we haven't. So, in compensation, precise technical ways of detecting colors and hues in wine have been developed. As a result, the science behind how light is absorbed in wine is quite advanced, and winemakers are beginning to pay attention to it.

There are three major components to a wine color. The first is intensity. This involves simple quantification of how dark the wine is as a result of absorbance of light at three different wavelengths. A sample of wine in a small glass container is placed in a spectrophotometer. This machine blasts light of specific wavelengths through the wine and measures the light that emerges. The amount of light that makes it through is proportional to the amount of light-absorbing chemicals (such as anthocyanins) in the wine. Light is sent through the wine around three points in the visible light spectrum: 420 micrometers (violet), 520 micrometers (green) and 620 micrometers (red). The wine color intensity is the sum of the absorbencies at these three wavelengths.

The second measure of visual wine quality is hue. This is a technical measure acquired by taking the absorbance measured at 420 micrometers and dividing it by the absorbance at 520 micrometers. This measures the ratio of violet to green matter in the wine, which experts think is important. The third and most commonly used measure for assessing color integrates absorbance data taken across a wide range of the spectrum incorporating three aspects of color. The first of these is clarity, or luminosity (L: the up and down axis in the figure), which measures how white or black the wine is. This term is scored on a scale of 0 to 100, and the wine is whiter if the L is closer to 100, and blacker if it is closer to 0. The other two axes shown in the figure are known as a and b. The a axis measures

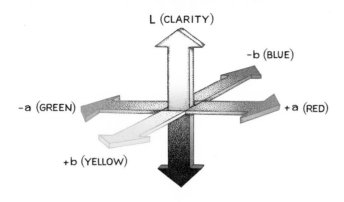

L (CLARITY)

−b (BLUE)

−a (GREEN)

+a (RED)

+b (YELLOW)

The color axes for wine

the redness or greenness of the wine (a positive value is red, and a negative value is green), and the b axis measures the yellowness or blueness of the wine (a positive value is yellow, and a negative value is blue). In this way, the spectrophotometer acts like an accurate all-seeing eye, rather like those tetrachromatic women. But the machine doesn't have the aesthetic reaction to the colors that the women presumably have.

So why would we want to know the color of a wine so exactly? For several reasons. First, the process of grape pressing and initial fermentation will have a huge impact on the overall color of a wine. Specifically, the amount of time the grape skins are in contact with the must affects the color of a wine, which in turn reflects how well the desired components of the wine have been extracted. This also has a direct effect on the fullness or body of a wine, so the winemaker can use color as a proxy for the heaviness or lightness of a wine. A wine in which color matches texture is a desirable commodity.

Winemakers have learned over the ages that color can also provide information on other important characteristics about the quality or texture or age of a wine. For example, the color of a wine is influenced by the amount of acid it contains. And for another, color is not a fixed attribute. As wines age, reactions occur among the various compounds and acids they contain: a red wine will usually evolve over time from a deep red to a tawny brown. White wines will tend to darken, until in really old wines it is sometimes difficult to know just from looking at a wine what its original color was. In addition, the vessel in which wine is aged may have an impact on the color, particularly if it is an oak barrel, which adds chemical complexity to the wine and affects its color as well as its aroma and taste. Finally, different blends of wines can be controlled precisely with careful color monitoring. And rosés benefit directly from the ability to assay color precisely.

✦ ✦ ✦

The sense of smell is crucial to appreciating one of the principal features of any decent wine. And there are good reasons why wine tasters routinely smell a wine before they taste it. Our sense of taste is limited (we have five basic tastes), whereas our sense of smell is complex. Smelling a wine before tasting it can enhance the variety of sensations that can be extracted from a good wine, and can help the wine taster discern immediately the differences between a good wine and an excellent one.

Before getting to the specifics of how our noses work, we first need to understand the classic ways in which the nose has been used to appreciate and describe wines. Three terms are commonly heard in this context, aroma, bouquet, and odor, and each has implications for wine. *Aroma* refers to the fragrances that emanate from wine as a result of its basic chemical makeup. *Bouquet,* on the other hand, is used to describe the scents that arise from the processes of fermentation and aging: the products of the wine's individual evolution. *Odor* is reserved for undesired smells and is usually meant to convey that something has gone wrong with a wine.

A developing wine is a potpourri of chemicals, such as sugars, phenols, and acids, and these can react with one another to produce new molecules. So the same wine will have a different smell at different stages in its development, although because most of these reactions occur early in the fermentation process, the major changes in aroma (and perhaps also in odor) will occur rapidly during this period. Changes in aroma slow down as fermentation proceeds.

The sense of smell is molecular, depending on the detection of particular molecules. And like all other molecules, those present in wine do not differ solely because of the different atoms of which they are composed. They also vary because those atoms are arranged differently, giving each molecule a characteristic size and shape. Imagine pouring a glass of wine. As soon as the bottle is uncorked, molecules from the wine start to float around in the air, though most stay close to the surface of the wine poured into the glass. There are billions of these airborne molecules, and there are hundreds of different kinds of them: alcohols, phenols, esters. Many of them float easily on the air because of their volatility, while others linger in the liquid and must be released by swirling the glass. At that point, our noses can begin to appreciate the complexity of the wine's molecular makeup. In many ways, the air above a wineglass is like a box of jigsaw puzzle pieces waiting to be assembled into a coherent picture. This process begins in the nose, which rapidly sorts out the kinds and relative amounts of the molecules present, information that is then rapidly and efficiently analyzed by the brain.

So how is this done? Go to the mirror, tilt your head up a bit, and look into your nose. With a little light you can make out its lining, otherwise known as the nasal epithelium. If you could zoom in microscopically, you would see that this is covered in little hairs, known as cilia. The cilia are swimming in a thin layer (0.06 millimeters) of mucus that efficiently traps the compounds that emanate from your wine, allowing the cilia to get to them quickly. Once cells on the surface of the cilia come into contact with the compounds, a chain reaction occurs that is broadly analogous to what happens in the eye, though it is less well understood.

There are two major schools of thought among smell researchers. One view holds that the nose works using a lock and key mechanism. The compound comes in contact with the cilial cells, which have odorant receptors embedded in their membranes. Just as in the retinal opsins, part of these receptors protrudes outside the cell. When the compound comes into contact with an odorant receptor that has the right keyhole, it binds to the receptor protein. This causes the protein to change its shape, inducing a chain reaction in the cilia cell resulting in an electric potential that is duly transmitted to an immediately adjacent part of the brain called the olfactory bulb. Neurons in the bulb then interpret the kind of smell indicated by the original compound. Like sight, what hits our brains is what we smell.

A second possibility is currently championed by the biophysicist Luca Turin. Instead of the lock-and-key mechanism, Turin contends that the compounds we can smell vibrate, and different compounds vibrate in different ways. The vibrations cause the odorant compound to transfer an electron to the receptor on the cell surface of the cilia, triggering a response in the receptor that starts the chain reaction eventually detected by our olfactory bulbs.

Whichever mechanism is the right one, the ability to discriminate between the many different odorant molecules that hit the nasal receptors comes from having a huge variety of receptors. With sight, there are only four different kinds of cone cells. But the human genome contains around nine hundred odorant receptor genes, found in the hundreds of different kinds of cilia in the nasal passage. And this is why our noses can interpret with such clarity the hundred or so different compounds that wines may give off, in a mind-boggling array of combinations.

✦ ✦ ✦

Asking, with the comedian George Carlin, "What wine goes with Cap'n Crunch?" might not actually be as trivial as it sounds. In fact, many people spend a lot of time worrying about which foods go best with which wines. This concern is not frivolous: the tastes in wine interact closely with all the other substances competing for the attention of our taste buds.

The process begins with the tongue. Take a glass of a nice, deep-red wine—for example, a young Cabernet Sauvignon. Sip it, let it bathe your tongue, and look in a mirror. Your tongue will resemble a field of little purple mushrooms or a mass of small purple pegs. These pegs are called fungiform papillae, and, though it's not apparent to the eye, they are not all alike. Each papilla is made up of between 50 and 150 cells. At the tip of each bunch of cells there is an opening called the taste pore. Little hairs (microvilli) protrude from this pore and come into contact with the molecules emanating from the substance we have put in our mouths. The microvilli are actually cells that bear receptor proteins for the molecules that convey taste.

Our sense of taste has fewer receptor types than does our sense of smell. There are five major kinds of taste: salt, sweet, bitter, umami (or savory), and sour. Bitter, sweet, and umami tastes are detected in the same general way as smells. Thus, items that taste bitter and sweet emit their own characteristic small molecules. These are shed from items placed into the mouth, and interact with the appropriate receptors on the microvilli. This causes a chain reaction in the interior of the cell that is transformed into an electrical impulse. In turn, this impulse is transmitted by nerve cells out of the receptor cell and to the brain. Salt and sour, on the other hand, are thought to operate through a different set of interactions. Instead of binding to a receptor protein, salty and sour molecules change the concentration of electrically charged ions, and hence change the action potential of the membranes in the microvilli. These action potentials are in turn sent to the brain for interpretation, just as the action potentials from sweet, umami, and bitter tastes are.

Not long ago, it was thought that four distinct regions of the tongue tasted different things. Thus, bitter-taste detection was believed to take place at the back of the tongue; sour at the middle of the tongue toward the sides; salty on the edges of the tongue toward the tip; and sweet at the tip. And this way of thinking about taste regions on the tongue led some wineglass makers to reengineer their glasses according to principles that are now in question. The claim was that different-shaped wineglasses delivered fluids to specific parts of the tongue or mouth, and thus to particular taste detectors. Accordingly, manufacturers touted their products as enhancing the taste of wine drunk from glasses specifically designed for Chardonnays, say, or for Cabernets. One company claims that one of its glasses directs wine to the center of the tongue, while another delivers to its tip. The corresponding recommendation is that the former should be used for wine of intermediate acidity, and the latter for wines of higher acidity.

There may be some truth to such claims. But two discoveries have conspired to cast doubt on them. First, molecular analysis of the different taste receptors in the tongue has debunked the notion that this organ is partitioned into taste regions. The field of fungiform papillae on the tongue is made up of a heterogeneous distribution of papillae that detect the five different tastes without reference to region. The brain doesn't care where on the tongue a food or beverage hits. Second, the fifth taste, umami, has thrown a wrench into the works. Umami is involved in how we taste small molecules called glutamates. These occur in wine, and yet the "region-specific theory" of taste on the tongue has no umami region.

Drawing of a tongue showing the rough surface of the tongue where taste receptors reside. One of the early theories of tongue taste receptors suggested that the receptors for four of the five kinds of taste were localized on the tongue. But this theory has been dropped.

How we combine the tastes and styles of wine and food has recently become big business, though basic commandments about combining food and wine have been around for decades. The first commandment is never to consume wine with garlic, spice, vinegar, or raw fruit. These foods tend to overwhelm the subtle taste of wine. And if you think about the receptors in your mouth that transmit taste sensations to your brain, the prohibition is easy to understand. Garlic, spice, vinegar, and raw fruit are all full of molecules that will react easily and strongly with the receptors of your tongue, leaving few receptors available to register the taste of wine. Something similar applies to other strong or greasy foods, such as pungent Stilton cheese or fatty foie gras. According to the second commandment, you will have to pick your wine carefully for these—sweeter wines, such as Port and Sauternes, respectively, are usually recommended. The third commandment is never to drink white wine with red meat or red wine with fish. That injunction was always on shakier ground, and nowadays, when it's not unusual for a chef to poach a filet of turbot in red wine, it is entirely negotiable. The basic challenge in drinking wine with food is to match the precise flavors and textures of the wines and foods involved.

So the real test is to determine the combinations of food and wine that will best enhance the taste of both. And understanding the five basic taste receptors and how to stimulate them will open the way to more enjoyable and appropriate combinations. So, for instance, if you are eating a salty dish, you might want to stay away from a wine that stimulates the salt receptors. But also keep in mind that if you eat salty food and drink a sweet wine, that wine will taste sweeter than usual. This is because the salty food will have blocked the majority of your salty-taste receptors, and the tongue will ignore the salt in the wine, even as it is detecting all the sweet molecules. Your available receptors will taste only sweet, which will thereby appear to be much more intense than if you had started with a clear palate. When eating sweet food, on the other hand, an acidic or heavy-textured wine might be best. Chefs have become creative in their wine and food pairings in recent decades, but whether we recognize them or not, successful pairings will invariably draw on the invisible chemical principles of our taste receptors.

✦ ✦ ✦

Glasses of different shape do direct fluids to different parts of the tongue, although it is not clear whether this attribute of the glass has much effect on the taste of the wine it contains. But where the wine hits the tongue is not the only purpose of differently shaped wineglasses. Even informal experimentation will prove that the glass enhances the sensory experience.

One of the most important features of a wineglass is its thickness, an attribute that underscores the importance of the sense of touch in the overall experience of drinking wine. A thick, clumsy, rounded rim will blunt the experience of drinking any wine, however expensively engraved the cut-glass receptacle may be. A thin, cleanly ground rim, on the other hand, will impart a degree of definition not obtainable any other way.

Also important is the size of the glass's bowl. Direct experience of a wine comes not only through taste but also through smell, and wines need space to oxidize and expand in the glass, and to release the aromatic and volatile molecules that will stimulate the olfactory receptors. Larger glasses have more than an edge here, especially for red wines. The glass also needs to be shaped in such a way that it can trap and concentrate the rising molecules for appreciation by the nose. It is more debatable whether red wineglasses need be the enormous balloons favored by some who attend commercial wine-tasting extravaganzas, or whether specific varietals need to be savored from glasses of different sizes and shapes.

But what we can affirm is that while compact receptacles such as the tulip-shaped 7¼-ounce (215 ml) official glass of the French Institut national des appellations d'origine et de la qualité may reduce clutter at tastings, and produce a level playing field, they are not big enough to bring out everything a wine has to offer. A better general solution is the similarly shaped but larger (12 ounce, or 350 ml) all-purpose glass in thin crystal available relatively inexpensively from several manufacturers. Some insist that red wines demand an even larger or more open bowl. With your own unique sensory anatomy, only you can decide by trial and error which glass—or range of glasses—is right for you. Fortunately, there are many choices.

One type of wine clearly demands a glass of specialized shape: the sparklers. In the past few decades the spill-prone open goblet, supposedly modeled on Marie-Antoinette's breast, that ensured the most rapid possible dissipation of the bubbles in Champagne and other sparkling wines, has been largely abandoned. The tall, narrow flute has eclipsed this cumbersome vessel; it provides maximum visual satisfaction in watching the bubbles rise, and it preserves them for a longer period of time. Still, elegant and compact as the flute is, it has its critics. An excessively narrow flute, it is argued, will dull the aromas of a sparkling wine and accentuate its acidity. Accordingly, many experts advocate tulip flutes, which are broader in the beam than regular flutes and have a somewhat wider opening. Our favorite is a broadish flute with a hollow stem, which lets the drinker enjoy watching the bubbles rise all the way from its foot.

As Gérard Liger-Belair reveals in his engaging *Uncorked: The Science of Champagne*, those bubbles—which are composed of carbon dioxide gas that had been in solution, under pressure, before the bottle was opened—require impurities on the glass in order to form. A bubble needs to nucleate in a vacuity that is at least 0.2 micrometers across, and nowadays, with advancing technology, manufacturing defects in the glass itself are typically smaller than this. So in theory, if the flute were perfectly clean, not a single bubble would form in Champagne. All the gas would escape directly into the atmosphere from the surface of the liquid, rather than rising in those mesmerizing bubble streams from below. Hurrah for imperfection!

✦ ✦ ✦

Having considered the senses we come to the brain, the hugely complex organ within which all that sensory information is processed and synthesized. We don't just taste with our senses, we taste with our minds. And our minds are routinely affected by a host of influences of which, quite often, we are not even aware. Both our senses and our common sense can be led astray by any number of extraneous factors originating in what we know, or think we know, about the wine we are drinking. Figuring out how our minds work in such complex domains as the evaluation of wines—which are, among other things, economic goods—is the province of neuroeconomics.

To study the relationship between consumer preference and, for example, the cost of wine, neuroeconomists typically set up blind experiments, in which the subjects are unaware of the parameters of the experiment. Researchers at the Stockholm School of Economics and Yale University recently conducted a double-blind experiment—in which both the subject and the experimenters with whom they come into contact are unaware of the parameters involved—upon this relationship. Their sample of over six thousand subjects included experts, casual wine drinkers, and novices. The experiment was simple. Subjects were asked to taste a succession of wines and rate them as Bad, Okay, Good, or Great. The wines ranged in price from $1.65 to $150, and the subjects were not told the cost. The responses for each wine were tabulated, and statistical analyses applied. Now, the average wine buyer might have hoped that this experiment would show that the price of a wine is correlated with its quality. This would certainly simplify life. But the researchers discovered that "the correlation between price and overall rating is small and *negative,* suggesting that individuals on average enjoy more expensive wines slightly *less.*"

To explore this relationship further, researchers at the California Institute of Technology set up an experiment in which they examined not only the dynamics of preference but also which regions of the brain might be

**Pricing Data in Caltech
Neuroeconomics Experiment**

Offering	Price	Price revealed to subject
1	$90	$90
2	$90	$10
3	$35	$35
4	$5	$5
5	$5	$45

controlling such preferences, in light of cost. To localize these, they turned to a technique known as functional magnetic resonance imaging (fMRI). The tough part of using this method on taste judgments is that the subject has to lie completely still, so the researchers had to devise a pump-and-tube system (a long way from those crystal glasses) to deliver the wine to their subjects. Then the researchers threw a complication into the study that allowed them to pinpoint whether knowledge of price affected perceptions of taste.

First, they bought Cabernet Sauvignon from three different vineyards: an expensive $90 bottle, an intermediate $35 bottle, and a rock-bottom $5 bottle. Their subjects were all young males (aged twenty-one to thirty) who both liked and occasionally drank red wine, but were not alcoholics. They placed the subjects in their MRI machine, connected the delivery hoses, and told them they were going to taste *five* different kinds of Cabernet Sauvignon. For each of the offerings the subjects were told the notional cost of the wine (as listed in the table), and then the wines were pumped into the subjects' mouths in a predetermined sequence, for a set amount of time. Subjects were then asked a series of questions designed to determine their preference for each of the "five" wines. The experiment confirmed that perceived wine cost was a heavy factor in choosing preferences. But the real revelation was that a region of the brain called the medial orbitofrontal cortex was hyperactive in every one of the subjects while he was making his choice. It seems that we all use the same part of the brain to make decisions about wine, at least when money is involved.

This experiment clearly showed that the subjects' preferences for the wines used in the study were strongly influenced by what they believed the wines had cost, and that this calculation was processed in a specific part of the brain. That's a start. But the subjects were relatively young and naive about wine tasting, and one might legitimately wonder whether an expert wine connoisseur would have been tricked in the same way. This experiment has not been performed yet, at least with an fMRI machine. But it seems likely from the literature that prior knowledge is a significant factor in most people's appreciation of a wine.

The psychologist Antonia Mantonakis and her colleagues looked at preconceived notions from another perspective. Before giving the subjects wine to taste, the researchers first planted in their subjects' minds either the notion that they had previously "loved the experience" of drinking wine or that they had "got sick" from it. Whether the subjects actually remembered their earlier drinking experiences in either way was irrelevant to the experiment, since virtually everyone has had experiences of both kinds at some time in their wine-drinking lives. What *was* important was the initial suggestion offered to the subjects. And the outcome was perhaps to be expected: people who were given the positive suggestion were more influenced by it in rating the wines than those who were received the negative one. Clearly, the tasters' responses were affected by extraneous factors, and the researchers concluded, logically enough, that if wine retailers wished to appeal to their customers' personal experiences of wine, they should try to call up the most pleasant possible associations.

Neuroeconomists have also been able to demonstrate by experiment something that has long been understood from anecdotal experience—namely, that our perception of wine is influenced not only by what is in the bottle but also by what we see on the label. Researchers in Barcelona and Paris conducted blind experiments in which they evaluated the role of the shape and color of the label in forming consumers' preferences for wines. Although both variables were significant in consumer choice, the colors of the labels were less important than their shapes, or the shapes printed on them. The most successful labels were brown, yellow, black, or green (or combinations thereof), with rectangular or hexagonal patterns. You might ask whether preconceived notions of cost might have affected the outcome of the experiment. But since the researchers also discovered that there was no correlation of cost with label preference, the experimenters felt confident that their conclusions were valid.

Does how much you know about wines in general influence how you perceive a specific wine? And what is the value of a name? To assess at least the first question (getting at the second would presumably have been too expensive), researchers gathered experts, moderately informed wine drinkers, and novices, and presented them with an advertising campaign for a particular wine, a Zinfandel, before the tasting. The variables in this case were the quality of the wine as assessed by external experts and the preferences of the subjects. In all cases, the experts were unswayed by the mock advertising campaign, while the novices were influenced by it in making their choices. But what was most interesting was the reaction of the moderately informed wine drinkers. These subjects chose the same wines as the experts if, before issuing their judgments, they were allowed to consider both the ad campaign and what they knew about wine. Given time to consider their choices, they were able to set their preference based on the quality of the wine. But if rushed and not allowed time to think, they turned in the same results as the novices.

The results of the initial experiment prompted the researchers to repeat it with only novice wine drinkers. But now, before the tasting began they educated their subjects for twenty-five minutes about wine and its quality. These novices turned in the same results as the moderately informed group had done in the first experiment; and in this case, too, the key factor in judging the quality of the wines correctly was allowing the subjects to think about what they had been told in the training session.

On one level, experiments like these show that advertisers are learning more and more about what influences our choices in wine, and that they are going to find ever-subtler ways to influence people to buy their products. Consumers thus need to be on guard, because it is clear that how one experiences a wine is affected by a host of factors, some of which might seem to be irrelevant. (Mantonakis and her colleague Bryan Galiffi even showed that consumers significantly tended to prefer the products of wineries with hard-to-pronounce names!) The good news in all of this, though, is that if you educate yourself on what constitutes a good wine and you use this knowledge as a standard when tasting a new wine, you will more often than not be able to judge its quality accurately.

✦ ✦ ✦

By the time you've swallowed a sip of wine, then, it will have engaged all five of your senses. In fact, a great wine is capable of delivering one of the richest multidimensional sensory experiences you will ever have— also, regrettably, one of the most expensive. Indeed, however you may score or describe the color, the clarity, the nose, the taste, and the mouth-feel of a wine, the end product will inevitably be summed up by just one number: the price. Although price and expectation go hand in hand, price and quality do not necessarily do so. It's a confusing market. So it's hardly surprising that a profession has grown up around the sensory evaluation of wine as an aid not only to its production, but to its consumption.

Once upon a time, the top wine critics were English. They were, by and large, aesthetes who celebrated wine as part of a much larger total experience of life. They tended to describe the wines they evaluated in relatively abstract and stylistic terms: a wine was aristocratic, lean, restrained, or voluptuous. Eventually they began ranking wines by awarding stars to them (usually between 1 and 5), and then, as the profession became a little more focused, by adopting a 1 to 20 scale. Those rankings were a bit like the 1855 Bordeaux classification described earlier: they had a tendency to reinforce a hierarchy that already existed.

Then came the Americans, led by Robert Parker. A lawyer by training, Parker started his career as the world's most influential wine critic by publishing a wine newsletter, and he became well known when he was faster than most of his rivals to single out 1982 as a classic vintage in Bordeaux. After this triumph, his *Wine Advocate* newsletter began to circulate widely in the trade.

Like his British counterparts, Parker carefully described the wines he rated, although he used a different vocabulary, based less on style than on a wine's immediate impact on the taste buds. Suddenly, wines were jammy or leathery; they tasted of herbs, olives, cherries, and cigar boxes. But the most important ingredient of Parker's formula was to rate wines on a scale of 50 to 100, exactly as his readers had themselves been rated for their performance in high school. No wine could score below 50, and be-tween 50 and 60 a wine barely rated mention. A wine that scored between 70 and 79 was merely average; it had to score in the high 80s to merit serious attention. Here was a scale with which all Parker's readers could identify, and although detractors railed (correctly) that such a finely gradu-ated scale was ridiculous, there is no doubt that Parker has a highly dis-criminating palate and knows a good or interesting wine when he tastes it. What's more, when he established his newsletter he deliberately es-chewed commercial sponsorship, and he paid for all the wines he tested. This was not true of *Wine Spectator,* a magazine that, after a lean start on newsprint, today rivals the glossiest of glossies in its production values, driven by lavish advertising, principally of high-production wines in the mid-to-upper segment of the market. *Wine Spectator* uses Parker's 50 to 100 scale, recommending only wines scoring above 75. Unlike *Wine Advocate's* practice, however, *Wine Spectator's* wines were usually evaluated by com-mittee, at least until some of its leading lights became minor celebrities in the wine world, averaging the scores of several tasters.

The numeric scale gives wine ratings an aura of impartial objectivity. But, as human beings, Parker and the editors of *Wine Spectator* remain crea-tures of preference. Rating something as diverse as wine by such a system is a bit like asking someone to rate blues and yellows on the same pref-erence scale: it can be done, but where each color tone will score entirely depends on which appeals more to the viewer. Still, there is enough agree-ment on what makes a wine great, or better than another, that a several-point spread will usually mean something significant to most people.

So the Parker rating scale caught on quickly. No longer did the wine buyer have to decrypt a critic's lyrical description to decide whether he or she would actually like the wine described; now it was as simple as pick-ing a wine that Parker had rated over 90. In turn, this meant a huge surge in demand for the wines that Parker liked, and prices for them rose ac-cordingly.

Several years ago, as wines he had been accustomed to drinking regu-larly skyrocketed out of his financial reach, one of us rather sourly re-marked to a wine merchant that he, at least, must have been happy with the Parker-driven price rises, which had presumably increased his margins. "Not at all," he replied. "If Parker gives it over 90 I can't buy it, and if he gives it less, I can't sell it." In its way, this is just as sad as the remark once made to us at a dinner party by an excessively affluent guest who declared that he only drank "the greatest" wines. Life, he said, was too short to drink anything else. It turned out that what he meant by "greatest" was actually "highest-scoring" and "most expensive." Well, if ever a strategy were designed to cut people off from the captivating variety that is the most intellectually entertaining and sensually rewarding aspect of drinking wine, this must surely be it.

Parker has always preferred lush, powerful, in-your-face wines like those produced in the Rhône Valley or in the Merlot-dominated regions such as Pomerol and Saint-Émilion that lie to the east of Bordeaux. And so pervasive did his influence become that producers all over the world began to use the technologies available to them to produce alcoholic, fruit-forward wines that would score high on the Parker scale. Out the window went ideas of terroir, replaced by a search for the wine that would score a perfect 100 on the Parker scale. An analytic laboratory was even established in Sonoma that, for a fat fee, advises all comers on how to produce a Parker 90+ wine.

The world is not a static place, however, and the Internet has changed the rules of the game yet again, allowing a huge chorus of pundits a voice and simultaneously creating a more perfect market that has taken away much of the thrill of the chase. In what we can presumably take as a nod to the times, even Parker not long ago sold a stake in his newsletter to Singaporean interests. But there is no doubt that Robert Parker's precise attention to numbers and his detailed criticism caused fine winemakers worldwide to pay extra attention to both the growing of their grapes and their winery procedures, and it contributed to a general rise in standards that was also driven by improvements in technology.

However high those standards might have been, though, this dynamic also promoted a growing worldwide uniformity of style, leading many to lament the increasing "globalization" of tastes in wine. If you have not seen the movie *Mondovino,* do so: its production values may not be the greatest, but its message—that the soul of wine is being lost as an internationalized mass market develops—comes straight from the heart.

Another effect of globalization has been to turn certain varietals into stars, sidelining the rest—although some, such as the Galician Albariño and the Campanian Aglianico, are making a comeback in boutique circles. As recently as the 1950s, few vignerons outside Burgundy were growing Chardonnay or Pinot Noir grapes, and the bulk Chablis and Hearty Burgundies that then accounted for so much of California wine production had never seen a Chardonnay or Pinot Noir vine. But nowadays, if you browse the wine list of any decent restaurant you will almost certainly find both varietals on offer from a bewildering array of vineyards around the world, while you will probably look in vain for a Savagnin. Yet although Pinot Noir character always contrives to shine through even in inferior versions, Chardonnay is remarkably responsive to circumstances. In different hands and places it can produce entirely different wines, making Chardonnay in a sense the ideal globalized varietal.

And of course, the fact that to the high scorers went the high prices was not lost on winemakers. In a world in which technology made almost anything possible, and the high numbers and big money often went to alcoholic fruit bombs, the winemakers duly followed, in a process lucidly captured by Paul Lukacs in his excellent history *Inventing Wine*. But for every action there is an equal and opposite reaction, and it is a rare pendulum that swings only one way. Some knowledgeable commentators are beginning to predict a shift in wine drinkers' preferences toward leaner, less alcoholic, more elegant wines, in which the balance has shifted toward structure and away from the fruit. We won't be sorry to see such a shift take place, although we wonder how it will fit in with climate change, something we'll discuss in Chapter 12.

Those nostalgic for the days when even people of modest means could occasionally afford a top bottle of wine might not think any development entirely bad if it lessened demand for better wines, and consequently lowered their prices. But at the same time winemakers need an incentive to lavish on their product the labor and investment necessary to achieve optimum results. Standards have improved as the returns on making good wine have risen. Nostalgia aside, the average table wine of our youth wasn't a patch on its modern counterpart, and it is not so long since most wine was rather pitiable stuff, the main attraction of which was that it was relatively safe to drink and/or got you tipsy.

This is good news for the average wine drinker who has just started exploring wines, or who can contrive to forget what he or she used to drink. Just as well, because the most expensive and highly reputable wines are becoming more and more significant as investment vehicles. In an avaricious world in which lucrative returns on capital are becoming harder to find, not only extravagantly wealthy individuals but even major hedge funds are buying prestige wines for their appreciation potential. What this means in practice is that much of the production of the top wines may increasingly disappear straight from the château (to avoid fakery problems) into climate-controlled storage, where it is likely to stay, occasionally changing hands at auction, until it is well past its prime.

Unfortunate as this may sound to those who think of wine drinking as a conduit to some of the most refined pleasures in life, it is hardly more tragic than the collecting and serving of wines purely as prestige items. This is increasingly happening all around us, as top wines become fashion accessories used to impress instead of being appreciated for their intrinsic qualities. The trend is accelerating, as vast quantities of top wine flood into affluent new markets where wines have not traditionally been drunk or enjoyed, and where, as the neuroeconomists understand so well, a bottle is likely to be appreciated far more for its price and label than for its contents.

10

Voluntary Madness

The Physiological Effects of Wine

W ell, if Jen Kirkman could drink it, so could we. We sat eying a bottle of the Sonoma County Cabernet (an impressive 14.5 percent alcohol by volume) of which Kirkman had consumed a liter and a half before giving an account of Frederick Douglass and Abraham Lincoln on the television show Drunk History. We knew from watching the show that the wine had gotten her smashed; what we wanted to know now was whether the show's producers had given her something decent to drink before she appeared on camera. We opened our bottle, and we are gratified to report they had.

Starting as an Internet craze, and now on Comedy Central, the American television show Drunk History features comedians trying to describe a historical event after drinking an immoderate amount of alcohol. Jen Kirkman was the first comic to appear on the series. After imbibing two bottles of wine, Kirkman, eyes rolling, face flushed, words garbled, gives a lecture on Frederick Douglass, employing such descriptions of Civil War luminaries as "Lincoln wasn't a douchebag." At one point she suddenly has to lie down to overcome a bout of dizziness, but then she continues, a new glass of wine in hand. In a slurred voice, she confuses Richard Dreyfus with Frederick Douglass, and President Lincoln with President Clinton. At the point in which she is describing Lincoln's assassination, Kirkman turns to the camera and asks, "I didn't take my pants off did I?" Apparently she has started to feel a chill in her extremities. She ends her recital with "Now my head is shutting to sleep" and "I have a mental illness."

Jen Kirkman's performance on *Drunk History* is a perfect representation of someone in the throes of what the Roman philosopher Seneca, two thousand years ago, delicately called voluntary madness. She shows the classic effects of alcohol on the brain: dilated pupils, slurred words, dizziness, loss of memory, altered physiology, drowsiness, and, above all, the shedding of inhibitions. How we get drunk has been the subject of much research, and scientific understanding of voluntary madness has increased. The recent attention to drunkenness is not purely because dependence on ethanol can be a social scourge. It is also because, as we hope to show, it is a complex physiological phenomenon. The human body has evolved to tolerate many chemicals and compounds that come into it as a result of breathing, eating, or drinking. The challenges alcohol makes to our bodies are ones that we as a species have faced for a long time. Evolutionary history tells us that our remote ancestors had to cope with alcohol, too, as evidenced by the existence of genes for alcohol tolerance in insects, worms, and other vertebrates. So it is not alcohol itself that is the problem: a little of it can actually be good for you. But overindulgence can be harmful, even deadly.

✦ ✦ ✦

What happens when a person drinks too much alcohol? To help understand the phenomenon of drunkenness, let's follow a wine-derived ethanol molecule as it passes through the body to the brain, and examine how this simple little drug impacts the physiological systems along the way. While following this path, many ethanol molecules will fall by the wayside, but the particular molecule we are concerned with will be one of the billions that will eventually make the subject drunk as they hit the brain.

As a glass of good wine nears the mouth, it will give off a bouquet and an aroma (most of us equate the two, but to a professional wine taster, *aroma* refers to the scents that come from the grapes themselves and *bouquet* to the scents that arise as a result of the process of aging) as some of the wine dissipates into the air as a weak vapor. Along with the molecules that give wine its wonderful bouquet and aroma, this vapor contains ethanol molecules and their byproducts. As we've seen, any of the molecules for which humans have receptors will register in the brain as a particular smell. Humans do not have an ethanol receptor in their olfactory system, but they do have receptors for a byproduct of ethanol called acetaldehyde—so purely by association, the acetaldehyde scent will elicit thoughts of ethanol.

The sensation continues as the wine passes the lips. Ethanol can bind both to a sweet-taste receptor on the tongue called T1r3, and to a bitter-taste receptor called hTAS2R16. Some ethanol molecules will be sucked up by these taste receptors and interact with them to tell the brain that something both sweet and bitter has been ingested. Mutated versions of the genes for both these receptors cause a tolerance for the taste of ethanol in mice (for T1r3) and in humans (hTAS2R16), and some hTAS2R16 variants in human populations are associated with an increased risk for alcoholism. Conversely, because some versions of hTAS2R16 taste bitterness more intensely because their proteins better bind ethanol, the signal these receptors send to the brain is stronger, often causing aversion. This bitter-taste receptor is an excellent example of how a tiny change in a receptor molecule can radically change behavior with respect to ethanol. But first impressions are not everything, and there are other reasons for drinking wine that involve deeper levels of pleasure than how it tastes.

As we follow the ethanol molecule, we need to bear in mind that it is one of billions in a bottle of wine, and how any one of these molecules impacts on the body depends entirely on where it finds a receptor—a matter entirely of chance. But as more ethanol is drunk, more of it will find its way to the organ systems it affects, and the effect of a bottle of wine far exceeds the effect of a glass. Another important variable is the blood volume of the drinker, because the ethanol from wine eventually crosses the membranes of the digestive tract and enters the blood. Indeed, blood-alcohol content has become the standard by which society measures—and judges—the amount of ethanol someone has in his or her body at any given time.

The math of blood-alcohol content is straightforward. A blood-alcohol content of 0.1 percent means that one-tenth of one percent of blood volume, or one part per thousand, is ethanol. This blood-alcohol content would mean that the person was pretty drunk. The blood alcohol sweet spot, the point at which a person becomes pleasantly tipsy, has been estimated at between 0.030 and 0.059 percent—a bit under the legal driving limit of 0.080 percent in the United States, and under or at the legal limit in most European countries. The amount of ethanol needed to reach a particular blood-alcohol content depends on body weight: heavier people have more blood. The time over which the ethanol is consumed is also important, because blood-ethanol concentration decreases as the ethanol is absorbed, released via the urinary tract, and broken down in the liver. For instance, bulky males would have to drink two glasses of wine in a half hour to exceed the U.S. legal limit, whereas a petite female would need to drink less than one glass of wine over the same time period. And, of course, the higher the blood-alcohol content, the more severe will be the impact of ethanol on the body.

Once past the palate, the ethanol molecule next hits a part of the throat called the pharynx, before sliding into the esophagus. Both these regions are lined with mucus that is packed with proteins and enzymes that normally begin the digestive process. But since ethanol is a molecule with which the machinery of digestion cannot deal, it remains unaffected by the digestive enzymes and passes by—though not without consequences, because it is actually toxic to some of the enzymes in the mucous layer of the esophagus. Besides altering the esophageal mucus, ethanol may also seep into the glands that produce saliva, occasionally in concentrations high enough to damage them.

The molecule has now reached the bottom of the esophagus, where it encounters the esophageal sphincter, the gateway to the stomach. If working properly, the sphincter will let food and drink into the stomach but not back out. Large amounts of ethanol, however, may cause the lower esophageal sphincter to become lazy, resulting in a backwash of some of the stomach contents into the esophagus. This is what is experienced as acid reflux, or heartburn. If it avoids setting off the backwash, the ethanol molecule will slide into the stomach, where it encounters a new set of cells, enzymes, and challenges.

Once in the stomach, the molecule finds itself in contact with digestive enzymes, especially the one known as pepsin. It also encounters the small molecule known as hydrochloric acid, which the stomach produces in abundance after the ingestion of food. As a small molecule, ethanol is relatively impervious to the digestive enzymes that target the larger proteins. But it can damage the stomach by overstimulating the production of digestive enzymes in low doses, and by shutting down their production in high ones. Any amount of ethanol, however, will disrupt normal functioning. Food in the stomach will help by sopping up ethanol molecules and keeping them from doing too much damage, and it will also absorb the ethanol and prevent it from entering the bloodstream.

After the ethanol passes through the stomach, it finds itself in the intestinal tract. From the small intestine, the molecules pass across the intestinal lining and are absorbed into the bloodstream. But in both the small and large intestines, the ethanol in wine can continue to make mischief by causing the muscles of the intestinal walls to weaken and underperform, allowing food to pass through more rapidly than usual. This accounts for the diarrhea sometimes encountered after a drinking binge. At the risk of appearing indelicate, we might also mention that occasionally people report having green excreta after an evening spent drinking red wine. This apparent paradox is caused by the rapid passage through the weakened intestine of the green digestive enzyme known as bile.

Ethanol readily passes across the membranes of the small intestine and into the bloodstream, which transports it to other regions of the body. The molecule's first stops around the bloodstream are in the organs that further break down ingested nutrients, notably the kidney and the liver. In the words of Murray Epstein, a specialist in kidneys and how they work: "A cell's function depends not only on receiving a continuous supply of nutrients and eliminating metabolic waste products but also on the existence of stable physical and chemical conditions in the extracellular fluid bathing it." And when ethanol is in the extracellular fluid that bathes the cells of the kidneys, some interesting things happen. The kidneys regulate the levels in the body of water and of several electrolytes such as sodium, potassium, calcium, and phosphate. Abnormal concentrations of these electrolytes can wreak havoc, and may eventually cause loss of kidney function and even death. Ethanol is toxic to the release of the antidiuretic hormone vasopressin, and it also stimulates the kidneys to increase the amount of urine produced. When vasopressin is absent or inhibited, the intricate tubes in the kidney tend to release water, diluting the urine produced by the kidney. As a result, the electrolyte concentration in the bloodstream goes up, and the body senses dehydration. This is why it is a good idea to drink a lot of water when imbibing wine and other alcoholic beverages.

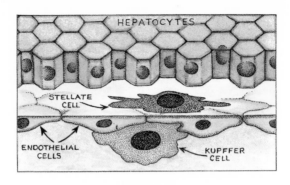

The cellular organization of the liver (the stellate and endothelial
cells play other roles in liver metabolism)

Now that the ethanol molecule has circumvented the actions of the
kidney, it must pass through another major organ of the body. This is the
liver, a massive organ (the largest in the body, weighing a little more than
the brain) that filters the blood. The liver is a fibrous mass made up of sub-
units called lobules, in which the filtering process occurs. There are more
than fifty thousand lobules in a healthy liver, each of which has several
veins running across it. Branching from these veins is a slew of smaller
capillaries that form a canal-like system leading to a central vein through
which the filtered blood exits. This system acts to increase the surface
area of the lobule, and hence improves the chances for the blood to come
into contact with the cells of the lobule. The canals are lined with cells of
two kinds. The Kupfer cells are immune system cells that eliminate bacteria and other large toxic items, while the hepatocytes, the workhorses
of the liver, do a broad array of jobs that include synthesizing cholesterol,
storing vitamins and sugars, and processing fats.

The liver function of importance here, though, is the metabolism of
ethanol arriving in the bloodstream. This process is dependent on an enzyme called alcohol dehydrogenase, or ADH, which converts ethanol into
acetaldehyde through oxidation (basically the reverse of the process of
fermentation described in Chapter 5). Acetaldehyde is extremely toxic to
the body, so many organisms, including humans, have evolved the ability
to rapidly break it down into useful acetate. The enzyme that undertakes
this detoxification is called aldehyde dehydrogenase, or ALDH, and it is

specified by two kinds of genes: ALDH1 and ALDH2. Acetate is valuable fuel for the body, so once the liver breaks down the ethanol-derived acetaldehyde, it is transported to other organs for further processing.

As hinted in Chapter 2, ADH and ALDH did not evolve initially to aid in detoxifying the body of ethanol. Over evolutionary time, our ancestors were probably not ingesting much of this chemical. Instead, these two enzymes were originally important in the metabolism of vitamin A (also known as retinol), and seem to have been pirated away for their function in ethanol metabolism. Their new dual function results because retinol and ethanol molecules have similar shapes, and it is the shape of the molecule that ALDH enzymes recognize.

Liver cells metabolize ethanol in another way, using an enzyme called cytochrome P4502E1 (CYP2E1), which also oxidizes ethanol to acetaldehyde. The liver does not usually have many CYP2E1 enzymes, but when it is chronically bombarded with ethanol it tends to produce more of them. Excessive CYP2E1 has been associated with cirrhosis, the scourge of the liver, a condition that occurs when normal metabolic processes of proteins, carbohydrates, and so forth are disrupted by alcohol. Eventually the tissue of the liver begins to atrophy, and the hepatocytes start to die. This leaves the liver riddled with Mallory bodies (damaged filaments inside liver cells), one of the telltale signs of cirrhosis, a disease for which there is no cure.

Assume that our hardy ethanol molecule has avoided breakdown into acetaldehyde in the liver, and is still in the blood system. Eventually it will arrive at the brain, where it makes its major immediate impact on behavior. Ethanol traverses cell membranes fairly easily because it is relatively small, an attribute that helps it cross the "blood-brain barrier." Once it gets to the brain, its main action is to interfere with the molecules, embedded in the membranes of neural cells, that are known as NMDA (N-methyl-D-aspartate) and GABA (gamma-aminobutyric acid) receptors.

The NMDA receptors are important in regions of the brain responsible for thinking, pleasure-seeking, and memory; like the sensory receptors for smell and taste, they bind to molecules that cause chain reactions in cells and thereby influence the transmission of information in the nervous system. The proper functioning of NMDA receptors is ensured by molecules known as glutamate receptor proteins, which interact with two kinds of small molecules: glutamate and glycine. Once these are bound to the right places in the NMDA receptor system, an ion channel opens. Proper brain activity depends on normal glutamate and glycine action.

In *The Astonishing Hypothesis*, the distinguished biochemist Francis Crick remarked, "A person's mental activities are entirely due to the behavior of nerve cells, glial cells, and the atoms, ions, and molecules that make them up and influence them." Crick was saying, in essence, that glutamates and glycines are the currency of how humans think and behave. Those neurotransmitters are crucial to our nervous system, accepting and rejecting small molecules based on the needs of each neural cell. But various small molecules that are toxic to the nervous system have developed several ways to circumvent the neurotransmitters. One way is that of the "competitive antagonists," which resemble the receptor proteins. But by far the most common method is that of the small molecules known as noncompetitive antagonists. These bind to other sites in the NMDA receptor proteins, changing their conformation so that the neurotransmitters don't work. And uncompetitive antagonists, as small molecules that clog up the channels, confuse neural communication as well.

Ethanol is considered a noncompetitive antagonist, along with other molecules such as ibogaine (a controlled substance); methoxetamine, a so-called designer drug that currently is not controlled; and gases such as nitrous oxide (laughing gas) and xenon. But ethanol does not affect the brain simply as a noncompetitive antagonist. It additionally acts on the GABA receptors. When these are hyperstimulated by heightened ethanol levels in the brain, the ion channels they guard stay open, allowing chlorine ions to collect on one side of the cell. This disrupts normal ion distribution in the brain, and the neural cells stop communicating with one another. Whichever the mechanism, the end product of increased ethanol concentration is malfunctioning receptors and abnormal communication among the brain cells.

If the ethanol molecule is persistent enough, it might be transported to one of three particularly important areas of the brain where the NMDA receptors are in high concentration. These are the cortex (where much of our thinking occurs); the hippocampus (responsible for mediating mem-

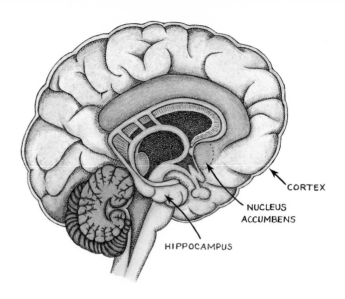

A cross-section of a human brain

ory); and the nucleus accumbens (from which reward-seeking behavior emanates). If the ethanol molecule binds to an NMDA receptor in any one of these regions of the brain, even though it is not near the glutamate or glycine binding sites, it will nonetheless change the shape of the protein, causing an alteration in the way glutamate binds that leaves the ion channel controlled by the receptor wide open. The open channel then stimulates that part of the brain, and a pleasurable sensation is felt.

The pleasurable feelings will continue even when the alcohol intake is excessive, but other side effects will develop. The high ethanol concentrations that reach the brain will desensitize the NMDA receptors, making them unresponsive to normal stimuli. Because the areas of the brain impacted in this way include both the thinking areas in the cortex and the pleasure areas in the nucleus accumbens, the higher functions will start to dwindle as more alcohol gets to the brain. Meanwhile, the ethanol molecules will also be affecting the other parts of the brain where GABA receptors reside. This happens as GABA receptors (and also some NMDA receptors) start to shut down in the hippocampus, a critical region for memory. After drinking two bottles of wine, Jen Kirkman's brain was in no condition to formulate a coherent narrative about Frederick Douglass.

If the ethanol molecule has been unable to bind to a receptor in the cortex, it might travel on to the occipital lobe of the brain. This region processes visual stimuli from the outside world. Ethanol has a toxic effect on the metabolism of glucose, a sugar that is an important source of energy for cells. If ethanol hits the occipital lobe in high concentrations, it will slow down the processing of glucose by some 30 percent. This means there will not be enough energy available to process accurately the images coming in from the eyes. The cells will cease to communicate properly, and visual problems will begin. Although double vision is perhaps the most famous visual impairment that results from too much alcohol, it is only one of them.

People frequenting the Internet chat room called "I'm Drunk and the Room Is Spinning" often complain of precisely the effect its title describes. This phenomenon has a medical name: positional alcohol nystagmus. And it is a disconcerting effect of ethanol that occurs in the head, but not entirely in the brain. Movement in many parts of the body can be impaired by ethanol, and dizziness starts out in the inner ear, where the sixth sense, or balance system, resides. Acting rather like a gyroscope, the organs of the inner ear sense the position of the body through tiny structures called the semicircular canals, tiny fluid-filled tubes oriented in each of the three axes of space. Associated with each canal is a conglomerate of cells called the cupula. This deflects as the head moves, and stimulates cells with small hairs on them that are connected to a nerve that leads to the brain, where the information supplied by the stimulated cells is interpreted as the body's position in space.

Any ethanol molecules that reach the ear via the bloodstream will bathe the cupula and distort its cells, placing them in continuous contact with the hair cells. The resulting stream of impulses makes it seem to the brain that the body is rotating. It accordingly tries to keep balance by making the visual systems spin a little, hence the sensation of the spinning head. When the drinker finally goes to sleep or—more likely at this point—passes out, the effect of ethanol on the cupula will eventually wear off. But sometimes the room still seems to spin when the drinker wakes up. Why? Well, one feature of waking up after heavy drinking is that the brain remembers what was experienced before the drinker fell asleep, and thinks the head is still turning. So it tries to correct by spinning the visual system in the opposite direction.

✦ ✦ ✦

We have yet to explore two of the most unpleasant aspects of excessive wine consumption: hangovers and alcoholism. The first is a short-lived and for the most part endurable hazard, while the second is a debilitating and often tragic disease. Ethanol in large amounts creates a physiological conundrum, and one of nature's less endearing reactions is the hangover. If only hangovers had a single cause, researchers might have found a way to circumvent or alleviate them, but they result from multiple causes, making them much harder to manage.

We have already discussed one aspect of a hangover—the sensation that the room is spinning. But excess ethanol can affect many parts of the body. For a start, the ethanol sucks up water, dehydrating the bodily systems and leading to a number of uncomfortable, dangerous, and occasionally deadly physiological outcomes that most commonly include dry mouth, nausea, and headache. Those who have experienced a hangover might find it hard to believe that brain tissues and cells do not themselves have pain receptors, but that is nonetheless the case. Headaches are thus aptly named, because the brain is not what hurts. It is the pain receptors of the head and neck that are impacted, and they present a diverse set of pain phenomena—over two hundred different kinds of headaches have been described.

One important cause of headaches is the dilation of blood vessels in the brain. In addition to dehydration, ethanol lowers glucose metabolism levels, and this exacerbates dilation. Blood vessels in this condition cannot properly transport blood around the brain, and altered blood flow manifests itself in pain, as neural receptors called nociceptors are stimulated and send information to the brain. The pounding headaches that result from drinking too much wine are due to the abnormal pressure that occurs in the dilated vessels with every pump of the heart. An equally unwelcome side effect of stimulating the nociceptor cells is nausea. Less extreme is the excessive sensitivity to sound and light that is typical of the "morning after the night before" syndrome. This happens as the depressing effect of ethanol on the brain cells wears off, and physical stimuli such as light and sound become amplified enough to overwhelm normal levels of perception.

We also have to remember that the effects of wines, most especially red wines, are not due only to the ethanol. Acetaldehyde, that problem byproduct of ethanol, is already present when the wine passes the lips, along with tannins and other chemicals in abundance from fermented seeds and stems. These too have an impact on headaches. In fact, some drinkers claim that hangovers are more severe from red wines than from whites precisely because of the extra chemical complexity of the wine, and particularly because of the impact the tannins have on human physiology.

So why do humans get drunk? In particular, why do they sometimes become addicted to alcohol? These questions have different answers, depending on whether they are considered as aspects of human physiology, mental states, free will, or, perhaps most important, genetics. As with any human behavioral disorder the genetic basis of alcoholism is complex, involving numerous genes and a strong environmental component. A person who is highly genetically predisposed to become dependent on ethanol might manage to avoid becoming an alcoholic because of social mores, behavioral modification, or some other cultural or social reason.

To understand why some succumb to alcoholism and others don't, let us look at a few rather striking single genes that have been implicated in alcoholism. The human body does have some means of processing ethanol, especially within the liver. The two enzymes that are particularly important, ADH and ALDH, have been studied in depth, and it is clear that there is considerable variation among human populations in the genes that control them. People of Asian ancestry, for example, specifically those whose ancestors were already living in the Far East by tens of thousands of years ago, tend to have a particular variant of the ALDH2 gene called ALDH2.2. Close to 40 percent of Asians today have the variant, whereas it is rare in people with European or African ancestry. The ALDH2.2 gene produces a protein that is partly inactive, and fails to break down acetaldehyde. The toxic acetaldehyde thus collects in the tissues, an effect that is often initially visible in a flushing of the face but that later manifests itself in an array of uncomfortable physiological reactions. As a result, people with this ALDH variant tend to stay away from overindulgence in ethanol, and alcoholism has a generally lower prevalence among them.

But one group of people of ancient Asian ancestry is an exception to this finding. About seventeen thousand years ago, some intrepid people living in East Asia decided to travel east. They walked from Siberia to the Bering Strait, where a land bridge was exposed. Either walking or following the coast in boats, they crossed into the North American continent and moved down the Pacific coast, occupying most of North and South America within less than five thousand years. With them came their genes, and it might seem to be a safe bet that the ALDH2.2 gene would be found in high frequency among Native Americans. But studies have shown that the ALDH2.2 gene variant does not occur in these Native American populations.

A variant of the CYP2E1 enzyme has also been implicated in alcohol avoidance, and this enzyme is particularly active in the brain. People with the variant are much more susceptible to ethanol, become tipsy more easily, and tend to stop after fewer drinks. Hence they show a reduced tendency to alcoholism because they generally stop drinking before toxic ethanol levels are reached and before they become physically dependent. What is most intriguing here is the mechanism involved. Although CYP2E1 can work like ADH and ALDH by oxidizing acetaldehyde to acetate, it can also metabolize ethanol in a process that produces free radicals. Researchers who work with the CYP2E1 variant that enhances the impact of ethanol suggest that the free radicals are doing something in the brain that is very different from our classical understanding of what ADH and ALDH do.

The neurobiology of alcoholism—what is happening in the brain during addiction to ethanol—can help to explain the propensity to alcoholism that some people show. Key here is that human brains and those of other mammals and vertebrates have evolved in such a way that they seek pleasure. Pleasure reinforces some of the most basic things we do in life, such as eating, drinking, playing, performing good deeds, and having sex. If those activities were not pleasurable, we almost certainly wouldn't do them as frequently as we do. Because pleasure is a crucial aspect of both individual survival and the success of the species, human bodies have evolved elaborate chemical methods to deliver stimuli for pleasure to the brain and to retain memories of that pleasure, so that people will seek more of it. Several parts of the brain and some complex neurochemistry are involved in those reward systems.

Three areas of the brain are particularly influenced by ethanol: the ventral tegmental area (VTA), the nucleus accumbens, and the frontal cortex. These also happen to be the three areas of the brain involved in the reward system. The technical name for the part of the reward system that is influenced by ethanol and other drugs is the mesolimbic dopamine system. This refers both to the group of brain structures that includes the VTA and the nucleus accumbens, and to the important neurotransmitter that is impacted by some drugs. Pleasure starts out as an impulse to the VTA, where dopamine is released. The dopamine then acts as a chemical messenger to activate the nucleus accumbens, which is implicated in motivation and reward-seeking. If there is a single "sweet spot" in the brain for pleasure, this is it. The more dopamine the nucleus accumbens receives, the more intense the pleasure will seem, and the more powerful the reward-seeking response will be.

Researchers have shown that neurons with GABA receptors extend into the reward pathway (the VTA and the nucleus accumbens). When ethanol bathes GABA receptors and causes them to malfunction, the impacted neurons in turn release dopamine and another neurotransmitter, endorphin. This latter is involved both in analgesia and in feelings of well-being; when lots of endorphins are present, the pain receptors become numbed. As the saying goes, we "feel no pain."

Ethanol's impact on the reward system differs a bit from that of other drugs. Cocaine and amphetamines make a good contrast. These compounds also affect the reward system via dopamine. But unlike ethanol, they alter the dopamine receptors directly: they are more direct in their intensity of addiction, and harder to break through. Equally devastating, every dopamine receptor in the brain is impacted by these drugs. Comparing the effects of different addictions is tough, but cocaine and amphetamine addictions are particularly nasty because the focus of these drugs on dopamine results in more intense addiction. Alcoholism, while also debilitating, falls into a different category because ethanol's effects are not focused on a single receptor. Instead, while ethanol's impact on dopamine receptors is localized to the nucleus accumbens and the reward system, it also impacts other receptors, such as GABA and NMDA, which are widely distributed throughout the brain. This is a significant difference, and one that makes alcoholism difficult to classify as equivalent to other addictions.

The genetics of alcoholism has been studied for decades, with procedures ranging from studies of twins to the more recently developed Genome Wide Association Studies (GWAS). Studies of twins use behavioral data from monozygotic (identical) and dizygotic (fraternal) twins to determine the degree to which a trait is heritable, while GWAS uses whole-genome sequence data to associate regions of the genome with a disorder. Because alcoholism is so complex, the results must often be interpreted with caution. But at this point it appears that genes are responsible for some 50 to 60 percent of the risk of developing alcoholism, which means that the environment has an almost equal impact. Furthermore, although several single gene variants are known to increase the risk of alcoholism, this is not the only thing that these genes control. Geneticists, genomicists, and behaviorists are in general agreement that alcoholism is a heterogeneous disease controlled by many genes, and that there is no one single kind of alcoholism. The alcoholism a researcher observes in an individual from Chicago might only slightly overlap in its genetic basis with the alcoholism seen in an individual from Detroit, or even from next door. Individual behaviors that might be involved in alcoholism include such things as impulsive and externalizing behaviors, relaxed inhibition, risk-seeking, and sensation-seeking, all of them also with complex genetic bases. The precise genetic basis for alcoholism remains largely a mystery.

✦ ✦ ✦

Looking over what we have just written, we found it just as difficult to banish fleeting thoughts of taking the Pledge as to resist pouring a hasty glass of wine. Humans, as we've already remarked, tend to take good ideas to extremes, and as in all other realms of human experience, there is a calculation to be made. It is a good idea to moderate the intake of any alcoholic beverage, including wine, not only to avoid the short-term repercussions of over-imbibing, but also to avoid long-term addiction to alcohol. Yet, as we celebrate throughout this book, wine has since the earliest times played a special role in human life, both as an emblem of civilization and as an enhancement of our experience of the world. There is, quite simply, nothing to replace it. We can offer no alternative to the standard exhortation: drink, responsibly.

11

Brave New World

Wine and Technology

W ine has always been a product of technology, and winemaking as we are familiar with it could not have been developed until ceramic expertise was harnessed for economic purposes during the New Stone Age. Today winemaking is a high-tech industry. But we were absurdly gratified recently to obtain a bottle of a wine that had been made in a simple clay vessel buried in the cool earth, just as was done with the very earliest wines about six to eight thousand years ago. Made in the eastern Caucasus from the venerable Rkatsiteli grape, this bone-dry pale-amber wine had an astonishing freshness to it, with a muted nutty fragrance that seemed to reach out across the millennia.

For centuries, winemakers were typically impoverished farmers who employed age-old methods using ancient equipment in gloomy cellars that were often shared with sheep, cattle, and geese. During the eighteenth century the general level of economic activity in the fabled wine-producing region of Burgundy was so low that entire villages would go into virtual hibernation during the winter. Even in more prosperous wine-exporting areas such as the Bordelais, winemakers tended to do business much as their predecessors had done since time immemorial: most grapes were made into wine purely for domestic consumption, or to sell in bulk to shippers. It was those merchants and other city dwellers, not the vine farmers, who built the impressive nineteenth-century chateaux that dot the countryside of the Médoc. In all but the most prestigious appellations, the wines themselves tended to be hit-or-miss, and were often oxidized,

or coarse, or thin and acidic. Apart from being alcoholic, their main benefit was that they were usually safe to drink.

During the twentieth century everything changed as modern technology
intruded, though the first material incursion of science into winemaking
had already occurred in the nineteenth century, when Louis Pasteur demonstrated that bacterial growth was responsible for many kinds of spoilage.
Previous advances in winemaking had been the result of trial and error, but
Pasteur preached that knowledge of process could serve as a guide to procedure. Earlier, the French polymath Antoine Lavoisier had figured out the
basic chemistry of fermentation, and the Italian scientist Adamo Fabroni
had shown that in wine this process is implemented by yeast; but it was
Pasteur who discovered in the 1860s that fermentation of wine involved
an ongoing interaction between the yeasts and the sugars in the must. He
also demonstrated that, despite the traditional addition of sulfur salts to
discourage this, the yeasts could be overwhelmed by rapidly multiplying
bacteria if too much oxygen was present. This led to his key message: to
produce good wine, as much oxygen as possible had to be excluded from
the process. One immediate result of Pasteur's dicta was a great advance in
the quality of sparkling wines, as winemakers regularly began to add sugar
and yeast after the primary fermentation to guarantee a second fermentation in the pressure-resistant bottle. But for the most part, the ravages of
phylloxera late in the nineteenth century postponed significant technological advances in wine production until well into the twentieth.

In the first half of the twentieth century the general quality of wines improved in all major wine-producing regions except the United States, where
Prohibition brought progress to a halt. The quality of the average wine was
enhanced largely through a general upgrading of vineyard and cellar management that was propelled, in part, by the establishment in several countries of appellation laws. These specified, for a particular region or *appellation,* and in greater or lesser detail, which grapes could legally be grown, how
vineyards were to be managed, and how the grapes could be vinified if the
wine was to bear the appellation's seal—and thus sell for a higher price. But
rather than spur innovation, these rules tended to entrench the best existing practices—and may still act as something of a brake upon innovation.

Significant new winemaking methods had to await the period following World War II, when professional schools of wine science finally began to exert widespread influence. In France, the science of oenology dates back to the establishment of the still-revered Jules-Émile Planchon's institute of vine science at the University of Montpellier in the 1870s. But during the past century, the country's most famous oenologist was Émile Peynaud, of the University of Bordeaux.

Peynaud, whose career spanned the first half of the twentieth century, was not just a laboratory investigator; he was a tireless adviser to numerous wine producers in the Bordelais and ultimately worldwide. He created the role of consulting oenologist (though emphatically not that of the celebrity "flying oenologist" cultivated by some of his students) that is crucial to the wine world today. An indefatigable experimenter, Peynaud believed that winemaking was not an art but a science, and he insisted that those he advised adopt new methods that, while expensive, resulted in many improvements. He started in the vineyard, where he vigorously advocated limiting the yield by pruning and by discarding excess or poorly ripening bunches of grapes. He required vignerons to carefully monitor the ripening process and to pick the grapes at the optimum moment. When he began his career, most grapes in the Bordelais were being picked far too early. This practice decreased the risk that problems due to weather would be encountered before harvest, but it also meant that many wines were weak, thin, and acidic. Peynaud changed all these protocols. In addition, he preached that only the best bunches should be used for the best wines. Grapes from a particular corner of the property, or from one vine rather than another, might be used to produce the estate's flagship product, while others might receive "second" wine status. Peynaud's approach involved laborious and expensive sorting of the grapes, but it paid off handsomely at the top end.

Peynaud was also active in the winery after the grapes had been crushed. He insisted on strict standards of hygiene to prevent bacterial growth, and he encouraged winemakers to replace the ancient oak barrels that had typically been used for aging wines. But above all, Peynaud believed in controlling the heat produced during fermentation. We noted in Chapter 6 that at excessively low temperatures yeast will become inactive and at high ones hyperactive, and Peynaud was adamant about maintaining ideal fermentation temperatures. His philosophy often mandated the abandonment of traditional large oak fermentation vats in favor of today's ubiquitous stainless steel fermenting tanks, individually equipped where necessary with cooling coils or jackets—a technology initially developed in the Champagne region. Typically, such temperature-controlled vats are used not only during fermentation itself but also beforehand to cool the must and afterward to provide a stable environment for the young wine.

One special advantage of such equipment has been to allow high-quality wines to be made in regions such as Algeria, and even parts of southern France, that would otherwise be too hot to produce anything but the traditional plonk. Another favorite strategy of Peynaud's, at least for certain wines, involved following the initial fermentation by an additional "malolactic" transformation. Implemented by inoculating the fermented must with bacteria that convert harsh malic acid in the wine to smoother lactic acid, this transformation comes at the expense of the wine's acid backbone, and sometimes of the aromatic compounds present. Occasionally such softening happens naturally, but because of the trade-offs involved, until Peynaud came along it had been mostly considered a problem. Nowadays, the virtues and vices of malolactic fermentation in specific circumstances remain a favorite subject of debate.

Peynaud's introduction of a scientific perspective into the ancient craft of winemaking resulted in a revolution in the quality of Bordeaux wines, with reverberations among winemakers around the world. The top clarets became better and more reliable year after year, as producers took his advice to heart, and the quality of the lesser wines from Bordeaux—and ultimately from elsewhere—rose along with them. In later years, this result was amplified by the effects of the commercial transformation of the wine industry in the latter half of the twentieth century.

Though Peynaud towered over his contemporaries in influence, if the United States had an oenological sage of equivalent stature over the post-war period, it was Peynaud's almost exact contemporary Maynard Amerine, a professor at the University of California, Davis. He, too, was an academic with his feet firmly in the outside world, consulting for numerous California wine producers as the industry began its renaissance following Prohibition and the tribulations of World War II. Amerine's particular

Émile Peynaud (*left*) and Maynard Amerine

specialty was climate and vines; before the war, working with his older colleague Albert Winkler he had already determined that the same vine stocks produced different wines in different places, and that the most critical variable appeared to be temperature. Regardless of varietal, cool-country grapes took longer to ripen. They were also leaner, more acidic, and more deeply colored and extracted. Grapes grown in warmer places ripened faster and had higher sugar contents. At the same time, certain varietals did better in particular temperature ranges than in others. Using what became known as the "Winkler Scale," Winkler and Amerine produced a map that indicated on the basis of temperature zones which varietals were best adapted to which regions of California. After the war, besides turning his interests toward the sensory perception of wine, explored in his *Wines: Their Sensory Evaluation*, Amerine energetically advised numerous up-and-coming California winemakers on where to plant which varieties, and trained many of the winemakers who eventually fanned out over the state to create the California wine industry as we know it today.

✦ ✦ ✦

The impact of all these wine scientists' efforts in the postwar period was twofold. Viticulturists grew more attentive to which varieties to grow given the locations of their vineyards, as well as to how best to grow and manage them. Winemakers, meanwhile, began to exert more control over the transition from grape to wine, carefully monitoring every stage of the process and intervening when they detected deviations from the desired course. As a result, in the early twenty-first century few wines are left to "make themselves." Every winemaker with pretentions to quality has at his or her disposal laboratory facilities for real-time tracking of everything that happens in the vineyard and the winery. Even before the vineyard is planted, vine growers determine the optimum spacing of the vines, depending on variety and conditions. With careful pruning, and sometimes culling, they reduce the amount of fruit on each stem to increase the plant's investment in each berry that remains. Viticulturists interested in making the most concentrated wine possible from their grapes might routinely discard a full third of the berries before allowing the rest to ripen. Foliage might be removed to increase the individual grape bunches' exposure to the sun. Ripening grapes are regularly monitored for sugar and phenolic levels, and at top vineyards picking may be done in stages, so that only completely mature grapes are harvested.

But today, if the best time for picking has been misjudged, or if bad weather threatens a harvest, technology will come to the rescue. To avoid weak coloration or flavors or aroma in the wine, or simply to increase the extract, the grapes can be "cold-soaked" under refrigeration before fermentation begins. Or reverse osmosis can help later in the process, by removing any volatile acidity or excess alcohol. Prior to fermentation, the reverse osmosis machine can also be used to eliminate from the must any excess water due to a rainy harvest—unless a vacuum evaporator has already achieved the same goal. Ultrafiltration will clarify the wine, and can also be used to remove oxidized phenolics. It will also take care of excessively bitter tannins, unless the winemaker prefers to damp those down by microoxygenation—a small amount of oxygen, in the right place at the right time, can actually help produce a soft and supple wine. Electrodialysis can adjust tartrate levels and acidity, and eliminate unwanted potassium. And so on. Before the advent of such modern technologies and practices, the best way to compensate for variations in grape quality from year to year was to adjust fermentation time and the blend of differently ripening varietals. Now all bets are off.

The armamentarium of interventions available to a winemaker nowadays is almost endless. Still, although technology is readily available to

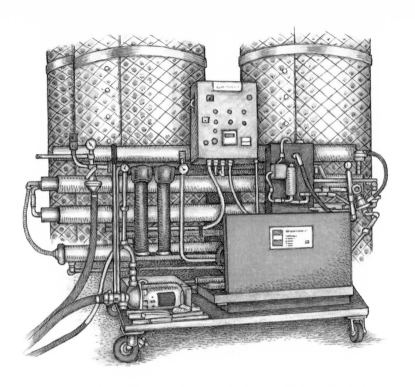

Reverse osmosis machine in a winery. Employing a sophisticated technique for filtering small molecules from newly fermented wine, machines like this one are often used to extract excess water due to rain near harvest time, to remove unwanted flavors, or to lower unduly high alcohol levels.

compensate for most problems encountered in viticulture, one thing remains unchanged: if the practices in the vineyard are right, less intervention will be needed to ensure that the wine turns out as the winemaker wishes. Nonetheless, at some point the wine has to mature, and even if the must is exactly as desired, there are many choices to make. The basic decision is whether to mature the wine in an inert material such as stainless steel or glass, with which it won't interact, or in wood, almost invariably oak. If the latter route is chosen, many varieties of oak are available, differing in tightness of grain, embedded compounds, and other variables that will affect the wine. Small barrels of around 225 liters capacity have become the norm, but winemakers must still choose between having barrel interiors that are strongly toasted or not. Coopers traditionally bend the staves of their barrels over a fire, and some flame-char the staves more than others do, a practice that will affect the barrels' influence on the wine. Another decision for winemakers is how often to change the barrels, because extract is lost with each usage. How long to leave the wine in the wood is also an issue: the longer the wine resides in the barrel, the more compounds it will absorb and the longer it will be exposed to tiny amounts of oxygen diffusing from the outside. Longer is not necessarily better—the choice is basically an aesthetic one—but if the winemaker is in a hurry, or the cost of new barrels is an issue, oak chips or worse can be added.

✦ ✦ ✦

Émile Peynaud was often attacked on the grounds that his scientific prescriptions amounted to an industrial formula for a soulless, standardized product. But in reality, this acute observer was exquisitely aware of terroir, as well as of the pitfalls of bad vine growing and winemaking, and he insisted that method be accommodated to place. What's more, his approach resulted in dramatic practical improvements: Peynaud's efforts unquestionably elevated the overall quality of wines made around the world, as much at the bottom end of the market as at the top. We are much better off for them. Still, Peynaud's insistence on rigor has produced a strong temptation among some to believe that there is an optimum procedure— and even an optimum product—in winemaking, an idea that was subsequently reinforced by the Parker generation of critics. It's certainly true that if the winemaker uses the formidable technology available to match an ideal set of parameters in the fermenting must, the wine will probably be good but is unlikely to be particularly interesting or innovative. This is one reason that, while table wines today tend to be of a much higher standard than their predecessors of half a century ago, they also tend to be more uniform.

Still, the scale of production also makes a huge difference, and it is remarkably how reliably large-scale winemaking operations can produce a standard product from grapes brought in from many different vineyards, sometimes kilometers apart, and make it consistent from bottle to bottle and year to year. For a large producer developing or maintaining a brand, this consistency is important. But it will come at the cost of nuance. Remarkably good wines have been produced on an industrial scale, but no great or truly exciting ones. The ones that really engage your attention are invariably produced in small lots, usually from specific places. Under such conditions the winemaker is able to handle each batch as the individual entity it is, and tailor its treatment to its specific characteristics. Providing this intensive care is not simply a function of the winemaker's intuitive genius and experience. It also requires the ability to charge enough for the wine to cover the high costs incurred without the economies of scale available to large producers.

Love the new approach or loathe it, adding climate science and chemistry to the traditional pursuits of wine production has inevitably shifted the balance between grape and terroir, and between terroir and control. As particular places became identified by the grape varietals that could or should be grown there, terroir as an abstraction has lost a bit of its mystique, while the grapes themselves have become to some extent subordinated to technological manipulation. For example, much to their chagrin, at a 1976 blind tasting in Paris a French panel of judges had extreme difficulty in discriminating top white Burgundies and red Bordeaux from Chardonnays and Cabernet Sauvignons produced in California. Even worse, a Cabernet from the Napa Valley took top honors. Paradoxically, the California winemakers had consciously been striving to emulate French models, which made their triumph something of a backhanded compliment to their cisatlantic colleagues. But the result was also evidence of a growing international convergence of style, something that was made possible by the increasing dominance of modern technology. Such developments engendered a wistful sense of loss among many, as commentators began to deplore the globalization of wine during the waning years of the twentieth century.

Naturally, there are still many mavericks who buck the trend: people like Josko Gravner of Friuli, on the border between Italy and Slovenia, who makes his highly regarded wines in huge clay pots buried in the ground, as his remote predecessors did in antiquity; or his neighbor Stanko Radikon who, emulating his grandfather in equipment if not in processes, macerates his white wines for several months in huge tapering oak vats. Both these radical departures from current best practices are taking place in the small town of Gorizia; and it is winemakers like Gravner and Radikon, in viticultural regions all over the world, who are making today's most inter-esting wines—albeit not always wines that are to everybody's taste, or even that their greatest fans would want to drink every day. Still, however unconventional they and their products may be, the majority of today's innovative winemakers are acutely conscious of terroir—of the patches of land to which they are both economically and emotionally attached. And their labors have shown that technical perfection in the making of wines can actually allow terroir to express itself to greatest advantage.

Nonetheless, some undeniably great wines, like the Grange Hermitage produced by Penfolds in Australia, deliberately shun the idea of terroir as defined by a specific place. The producers of Grange make a point of assem-bling the best Shiraz grapes from prized but widely separated vineyards, with the aim of allowing the character of the varietal to dominate (though small quantities of other grapes are sometimes also included nowadays to add balance where needed). Grange is a powerful and highly individual wine, unlike anything else, including top Shiraz-based but single-block wines made nearby, such as Henschke's almost equally legendary Hill of Grace. The message of both is that technological perfection need not get in the way of either terroir or varietal character. Idiosyncrasy of place, or of grape, need not be lost simply because the winemaker has gone to great lengths to avoid procedural mistakes.

At the other end of the spectrum, the endless meddling in the wine production process that technology now facilitates has made possible the marketer's dream: a global standardization of wine in a style appealing to a mass clientele. It is a triumph of technology that so much decent, drink-able wine can be produced on the vast scale on which the wine industry now operates, especially when compared to the seas of plonk produced in the past. But while those workmanlike wines are easy to enjoy, especially with food, they are far from what the very best handcrafted wines, grown in specific situations and individualistically vinified, have to offer. A good wine and a perfect one are worlds apart, distinguished not only by terroir and labor-intensive attention but also by a high degree of artistry. None-theless, those of us who love the element of surprise in wine have much to hope for as science advances. Science is no enemy of originality and sub-tlety in wines; as long as it is used to enhance the quality of the original grapes, rather than to disguise problems originating in the vineyard, we can hope for further serendipitous discoveries.

✦ ✦ ✦

No account of wine's place in science or human experience can avoid the subject of fraud. For, as long as certain wines have been judged to be superior to others, there have been fakes. Greeks and Romans frequently complained about the manipulation and mislabeling of wines. Pliny the Elder, for one, was vexed by a superabundance of fraudulent wine: at one time there was so much Falernian sloshing around in ancient Rome that most of it couldn't possibly have been genuine. In the fourteenth century Chaucer urged caution on wine buyers, especially when purchasing Spanish products, while during his sojourn in Paris the claret-loving Thomas Jefferson rapidly learned the wisdom of buying directly from the vineyard instead of trusting to the wiles of wine merchants. Indeed, it is to Jefferson's time that we may trace the origins of what we might call the modern period of wine fraud. By the late eighteenth century, when Jefferson was developing his buyer's instincts, winemakers had started using easily stacked cylindrical glass wine bottles, sealed with corks. Correspondingly, British aristocrats in particular had begun the practice of laying down bottles of long-lived wines—top clarets, Madeiras, Ports, Hocks, some Burgundies—for later consumption: a tradition that had begun after it was discovered that some wines continued to develop complexity in the bottle.

This evolution of the wine occurred as the oxygen in the air trapped below the cork—and to a minor extent diffusing through it—interacted with the compounds in the wine. Tough, highly extracted wines such as those produced in Bordeaux benefited particularly from being cellared, as the alcohols and acids in them softened and the tannins began to separate out. Although such wines were initially intended for later consumption by the purchaser or his descendants—you laid wines down for your grandson, even as you drank your grandfather's—the new form of packaging eventually created the conditions for a secondary market to develop, as bottles of older wines became rarer and more valuable. This was when the labels started to become as precious as what was in the bottle itself. In a traditional winemaker's cellar bottles were—and are—stored unlabeled, identified purely by the location of the bins in which they reside. Hence the presumably apocryphal, but certainly venerable, story of the cellar boy who came running up the stairs crying in a panic, "Master! Master! The cellar has flooded and the bottles are floating everywhere!" at which the cellar master calmly smiled and said, "No need for alarm, young lad. The labels are all safe and dry up here in my desk!" There was no such luck in a New York City wine warehouse during Hurricane Sandy, in which flooding has led to litigation that is likely to run on for some time—especially because it turned out that the wines were not insured. But the lesson is clear: from time immemorial there have been opportunities for substitutions and malfeasances along the supply chain leading from producer to consumer. And the invention of the bottle offered a range of new possibilities.

In the mid-twentieth century, two trends intersected. First, many members of the postwar aristocracy found themselves under financial stress but in possession of a lot of old wine. This fact did not escape the attention of auctioneers, who energetically cultivated the second trend, in which older top clarets and other age-worthy wines, preferably pre-phylloxera, were becoming increasingly sought after by collectors. During the 1960s appetite for these wines grew enormously, as reflected in skyrocketing prices at auction—despite the growing probability that the wines would be well past their prime because, no matter how hard and tannic a wine is to begin with, or how beautifully it may evolve in the bottle, eventually age takes a toll. Wines do not live forever. But in a larcenous world skyrocketing prices for what were basically prestige items, not necessarily destined to be drunk, made deception more profitable and less likely to be detected, and some notable scams began to stand out against the more usual minor fraud.

Thomas Jefferson himself figured in the most notorious recent scandal, entertainingly chronicled by Benjamin Wallace in *The Billionaire's Vinegar*. During the middle 1980s, a few ancient-looking bottles of wine from the Bordelais began to appear at auction and at select tastings attended by a coterie of well-to-do wine fanatics. What made these bottles remarkable was not just their age—they were marked with the years 1784 or 1787—but that the initials "Th. J." were also engraved on them. Hardy Rodenstock, the German collector who put them up for auction, claimed not only that they

One of Hardy Rodenstock's "Th. J." wine bottles, sold at
auction to Malcolm Forbes and never drunk

were part of a cache discovered in a walled-off cellar during the demolition of a house in Paris, but that the initials stood for "Thomas Jefferson." Here, by implication, were bottles that had been destined for Jefferson but had not reached him by the time he left Paris for the United States in 1789. Although Jefferson experts at Monticello declined to authenticate the bottles, the claimed provenance was largely responsible for the first of them, a 1787 Lafite, selling in London, at the end of 1985, for the equivalent of $156,000: four times as much as had ever been spent previously on a bottle of wine. The buyer was the American publishing magnate Malcolm Forbes, who put the wine on public display under hot exhibition lights, with the well-publicized result that the cork shrank and fell into the wine. Forbes had apparently not planned to drink the wine, but after this unfortunate incident nobody will ever know what that putatively ancient—and certainly monetarily precious—liquid tasted like.

With great fanfare, Rodenstock next produced a 1787 Mouton from the Jefferson cache at a private tasting held, in 1986, at the chateau itself. All in attendance pronounced it a great wine, still drinking well and developing in the glass. After this triumph, Rodenstock returned to the auction circuit with a 1784 Chateau d'Yquem, a Sauternes that, as an intensely sweet wine, had in principle the best chance of the lot of drinking well after two centuries. And although the authenticity of the Lafite sold earlier remained a matter for dispute, the Yquem went for a hefty $57,000. Now fast forward, via more sales and tastings, to a "supertasting" of 115 vintages of Lafite held in New Orleans in 1988, and for which Rodenstock supplied a bottle of the 1784. The wine was judged a disaster. Reportedly it was not simply oxidized, as were some other old Lafites; rather, it was *qualitatively* different. It was dark and acidic, and it didn't gracefully die in the glass, as an oxidized great wine would be expected to do. Many were puzzled, but the unfazed Rodenstock shortly afterward privately sold four more bottles from the Jefferson cache, plus some other eighteenth-century wines, to the hugely wealthy collector Bill Koch. The total value of the deal, completed through intermediaries, may have been close to $400,000.

As time passed Rodenstock's wine business blossomed, his discoveries of old wines became if anything even more extraordinary, and the tastings he supplied became increasingly rarified and extravagant. At the same time, doubts about the authenticity of his wines grew in some quarters, perhaps contributing to a general decline in the prices paid for old wines at auction as the 1980s came to an end. Eventually an appraisal of one private cellar that had been largely sourced from Rodenstock revealed a high proportion of probable fakes. These included one of the Jefferson bottles — a 1787 Lafite — that was sent to a laboratory for testing. While the sediment in the bottle was shown to have characteristics compatible with a two-hundred-year-old wine, the liquid itself yielded tritium and radiocarbon levels suggesting a much later origin, in the 1960s or 1970s. After these findings, the wine's German owner was prepared to concede it was a fake, a new wine in an old bottle that had retained its sediment. But even as legal actions and counteractions were launched, and more scientific testing was done, Rodenstock himself continued to prosper.

A significant change occurred in 2005, when Bill Koch began to have doubts about his four Jefferson bottles. He hired a private investigator, who discovered that the engraving on the bottles had been done with a modern dental drill. Other apparently incriminating circumstantial evidence emerged, and Koch brought suit directly against Rodenstock in New York City. The suit was thrown out in 2008 because the court decided it did not have jurisdiction; but by then the gig was up. Both Rodenstock and many luminaries of the wine world found themselves discredited, and even now both the legal and the sensory ramifications of the saga are far from being sorted out.

As Wallace emphasizes in his book, the key to such sorry tales, as so often in confidence trickery, lies in the willing collaboration of scammer and victims. Ignoring warning signs that had been available even before the first Jefferson bottle was sold, some of the most distinguished palates in the wine world had pronounced the fakes to be outstanding wines, elegant and durable representatives of their improbably remote period. Clearly, far too often the story was not the wines themselves but what their drinkers wanted them to be. The human brain is a mysterious organ, making connections that may or may not accurately reflect reality.

But the Rodenstock affair is only the best-publicized and longest-running of many such sagas. As long as collectors are prepared to pay enormous prices for rare bottles of wine, there will be those willing to supply them, whether genuine or not. Opportunities for fraud abound, particularly among top wines from the later decades of the twentieth century that continue to soar in value, and are contained in undistinguished, machine-made bottles. To make it even easier for counterfeiters at the highest end, rumors abound that there are nowadays precise recipes available for faking almost any kind of wine. Even simpler, though, it is also possible to give an old wine a new identity with a sought-after label, either forged or recovered from an old bottle. Indeed, it is nothing short of amazing how cavalierly the forging of labels has been done: sometimes bogus labels can be identified simply from spelling errors. Still, a few top producers are now guarding against forgery by incorporating markers into their labels, much as banknote printers do, and by using stronger glue to attach them to the bottles. Bottles are also being made more distinctive.

One of the most recent old-wine scandals has Rodenstockian resonances—though in financial terms the ante has by now been considerably upped. In 2003 Rudy Kurniawan, an Indonesian wine connoisseur with a reputedly awesome palate, burst onto the southern Californian megatasting scene. He quickly contrived to establish an almost gurulike presence among the nation's elite wine consumers, even as he entered the auction market, initially as a conspicuous buyer of rare wines (which helped drive up prices) and later as a major consigner of sought-after wines from Burgundy and Bordeaux. At two New York auctions in 2006, the contents of a cellar widely believed to have been his grossed a world-shattering, and frankly ludicrous, total of over $35 million. Once the shock had worn off, the sheer superabundance of rare wines at these auctions raised questions; by 2009, doubts hovered over the authenticity of any wine that had passed through Kurniawan's hands. The Indonesian had by this time stopped taking even elementary precautions, consigning to auction several lots of a wine that had not begun production until several years after the vintage indicated on the bottles. So blatant was this fraud that the proprietor of the vineyard concerned felt compelled to be present at the auction to ensure that the bottles claimed to be from his grapes were not sold.

Early in 2012, FBI agents raided Kurniawan's home in a Los Angeles suburb and found a trove of what they described as wine-counterfeiting paraphernalia. Evidence included a corking device, numerous foil capsules, hundreds of fake labels for wines dating back to the nineteenth century, and records indicating the purchase of large quantities of inexpensive Burgundy wine. In December 2013 Kurniawan was found guilty on two counts of mail fraud, and in 2014 he was sentenced to ten years in jail. Meanwhile, the world of super-grandiose wine-collecting and flashy wine-swilling once again has egg on its face, while—inevitably—the Kurniawan episode has been supplanted by newer scandals.

But before we conclude that wine counterfeiting is a problem encountered only by the rich and ostentatious, consider other relatively recent instances of wine fakery that have affected less prominent people. In 1985, Austria's export wine industry almost went under after it was discovered that a few wine producers had added small quantities of diethylene glycol (used in many brands of automobile antifreeze) to their white wines. They had taken this step to boost the apparent sweetness and body of otherwise thin and acidic wines, and thus make them more appealing to unsophisticated wine drinkers in neighboring Germany. No wine drinkers were harmed by the ethylene, and the consequent restructuring of its wine industry ultimately led to a huge increase in the quality of wines produced in Austria. But the next year, in Italy, at least twenty people died after drinking cheap wines that had been adulterated with methanol to raise their alcohol levels.

Even the most distinguished winemaking regions are not immune. In 1973, just as the fine-wine boom was beginning, the long-established Bordeaux firm of Cruse & Fils Frères, proprietor of the Cinquième Cru Château Pontet-Canet, found itself caught up in a scandal that involved falsified records which allowed ordinary red table wines to be mislabeled as Appellation Contrôlée Bordeaux. Although not personally concerned in the scandal, the family patriarch, Herman Cruse, committed suicide, the business was disgraced, and the château was lost, while consumers discovered that they had paid high prices for an inferior product. By 1998, when prices had increased yet more, another prominent claret producer, the Troisième Cru Château Giscours in the heart of Margaux, had also become embroiled in scandal. The property was accused of diluting its second wine of the 1995 vintage with wines from different years and regions, and of adding adulterants such as milk and fruit acids. Two employees were indicted, but the result of the litigation was never made public. Despite such episodes, Bordeaux prices have continued their inexorable rise.

One of the latest furors has hit Bordeaux's great rival region, Burgundy. In mid-2012 it was announced that the French authorities had begun investigating one of Burgundy's largest wine producers and shippers, the house of Labouré-Roi, on charges of fraud involving a staggering 1.5 million bottles of wine destined to be sold to consumers all over the world. Charges included blending in wines from external appellations, topping up fermenting musts from highly reputed vineyards with cheap table wines, and mislabeling. It is significant that the alleged fraud was discovered not through sensory evaluations, but by an audit that revealed that the physical volume of Labouré-Roi's wines had remained constant despite the expected losses to evaporation. Still, the large scale of the reported scam was hardly unprecedented, even for a prestigious appellation; in 2008 it was alleged in the Italian press that possibly millions of liters of wine marketed under the expensive Brunello di Montalcino label had not been made from 100 percent Sangiovese grapes as required by law, but were cut with cheaper grapes brought in from outside Montalcino. This "Brunellopoli" scandal reverberated so widely that the U.S. government threatened to ban all Brunello imports that could not be proven by laboratory testing to be pure Sangiovese.

The threat was not an empty one, because at least in theory it is possible to test for both locality and varietal. Each vine-growing region has a unique stable isotopic makeup for carbon, oxygen, and hydrogen, and a database of isotopic ratios exists for many regions of the world where food is produced. Because the isotopic composition survives the winemaking process, any wine can be tested for its general area of origin. Authentication to varietal is also now possible via two techniques based on DNA. One involves the infusion of DNA from the vines themselves into the ink that is used to print the labels on wine bottles. Some Australian wineries are using this method to track and authenticate their products. The other approach involves directly isolating DNA from the wine; fermentation is one of many food-processing methods that do not destroy DNA in the final product. Once the DNA of the wine in a bottle is isolated and sequenced, the sequence can be compared to others in the grape studbook to determine the vine strain used to make the wine. This approach has been utilized for years to identify caviar, some of which comes from endangered sturgeon species. With the right equipment, the DNA from processed caviar can easily be isolated and compared with a database of fish sequences to identify the species from which it came. In all such approaches, the DNA from the source of the wine or the foodstuff is used as a "barcode" to identify the species or variety of origin.

The authentication methods we have just described are new, and hold promise for the future. But the lesson of history is clear. At the top end, a lot more of many prestigious wines has probably been drunk than was ever produced; at the low end, the threat of adulteration of cheaper wines with cattle blood, battery acid, and other unpleasant additives is unlikely to go away, particularly if climate change increasingly threatens producers with poor growing conditions. Of course, those buying old and rare bottles have always been taking a risk, and many of the world's leading restaurants warn their customers that if they order a rare wine beyond a certain age there will be no swirling, sniffing, and sipping followed by sending it back. That is probably as it should be, certainly as regards the condition of the wine. Someone has to bear the risk. And while guaranteeing the wine's origin is a different matter, at this market level the caveat emptor rule seems fair. What is more, for those in search of the genuine grandes dames of wine, the Internet already overflows with more or less useful advice on how to detect a faked old bottle.

For everyday consumers, it will be business as usual. As long as there is a buck to be made by adulterating wines, someone will be doing it. If we want to be sure that we are drinking what we think we are, at present we are largely dependent either on our own knowledge and sensory evaluations or on official vigilance. Despite the promise of technology, in an age of deregulation it seems unlikely that the authorities will in future be more helpful than they have been in the past. For the most part, consumers will continue to be on their own—unless, of course, someone develops a wine-testing app: a discreet little probe, connected to a vast database via smartphone.

12

Franken-Vines and Climate Change

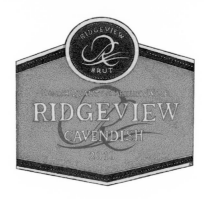

The Romans made wine throughout England, almost up to the Scottish borders. With a deterioration in climate vine growing subsequently dwindled, even on southern English soils that were almost identical to those of Champagne. During the twentieth century only a few hardy eccentrics dared make wine in such marginal conditions, but with renewed climatic warming, this is changing. We chose one from the increasing range of English sparkling wines now available. Made with the classic Champagne grape varieties, it was a revelation: fresh, vibrant, with a fine mousse and a hint of warm bread in the finish. We wished we could have afforded to buy a few more.

We started this book by peering into the ancient past; as we near its end, our focus shifts to the future. Looking back, we can see that the climatic history of the world has been notoriously unstable. At one time, the entire planet Earth was a frozen snowball; at another, dinosaurs lived in Antarctica. Asteroid impacts, variations in the planet's orbit around the sun, large-scale volcanism, and changes in the composition of the atmosphere are only a few of the influences that have interacted to produce enormous fluctuations in climates and environments around the globe. Going forward, conditions are unlikely to remain static for the wine industry, whatever the specific reasons for change at any particular time. We have glanced at the impact of modern technology on the quality of the wines we drink; now, as significant climate change in the near term looms as an increasingly alarming possibility, it is natural to ask how technology might help mitigate its effects on the vineyards and their products. One obvious candidate is genetic engineering.

Any human intervention that involves genetics—whether altering genes, directing particular kinds of interbreeding, or moving genes from one genome to another—is considered genetic engineering. By this broad criterion, humans have genetically engineered grapes for millennia, an activity that has played a large part in creating the broad spectrum of wines available. But traditional methods of genetic engineering are both imprecise and time-consuming. They can also be frustrating, especially when the desired product does not result from a cross. Modern gene engineering techniques, however, can alleviate both the wait and the frustration. As we saw, a genetic studbook for all the vines used in the production of wine has been constructed, using the genome sequences of thousands of rootstocks and grape varieties. Not only is this studbook important for tracking and identifying the different varieties of grapes involved in winemaking; it can also potentially be used to identify which crosses will yield desired qualities.

The grapevine genome has over twenty thousand genes. Some of these are essential to the vine's existence at the cellular level. Other genes are important to traits of the grape and vine, such as seed production and grape color, that make the wine tasty or in some way unique. To date, few of the genetic markers in the vine studbook have been connected to genes for specific functions. Rather, they have been identified simply because they vary among different grape varieties. But once a gene has been discovered that is involved in a specific wine characteristic, it is a simple task to determine which of the studbook markers is in close proximity—what geneticists call linked—to that gene. If, for instance, a genetic marker is shown to be linked to more efficient sugar production, or to a particularly attractive grape color, the vineyard manager can scan the studbook for varieties of grapes that have that trait coded in their genomes and make the appropriate crosses. The availability of the studbook allows the grape grower to become a more efficient matchmaker. As it is refined, its potential for modifying grape lineages will be enhanced. Thus, although grape growers two centuries from now will be doing the same basic job as their predecessors did thousands of years ago, they will have a much better grasp of how genetic crosses can enhance particular grape traits useful to wine production.

The molecular, or genomic, version of genetic engineering involves moving desired genes from one genome to another. Some of the more famous cases of agricultural molecular engineering include genetically modified corn that resists insect infestation and grows faster. This kind of intervention is controversial, but it is potentially valuable in two areas of plant and animal husbandry: the prevention of disease and the enhancement of traits that increase yield or quality. For grapevines, genetic engineering might be used to enhance traits relating to the palatability, purity, or alcohol content of the wine. Genetic engineering can also be used to ensure that vines and grapes are resistant to infections from insects like phylloxera, or bad fungi such as bunch rot. Indeed, as of 2005 there were already about thirty genetically engineered grape varieties in existence. The production of genomically modified grape varieties has slowed a bit in the past few years, but available varieties include forms engineered to enhance resistance to viral, bacterial, and fungal infections such as *Agrobacterium, Botrytis, Clostridium,* nepovirus (nematode-transmitted viruses), and beet yellow virus. In 2002, researchers at the University of Illinois in Champaign-Urbana, focusing on how grapes succumb terribly to the herbicide known as 2,4-D, transferred into a grape genome a gene from a soil bacterium (*Ralstonia eutrophus*) that degrades the chemical. Specifically, it was inserted into the Chancellor grape, yielding a grape strain known as "improved Chancellor."

Researchers at Cornell University have field-tested Californian vines into which they incorporated genes from the soil fungus *Trichoderma harzianum* in the hope of producing vines resistant to botrytis and powdery mildew, while Australian scientists have inserted a gene that prevents fruit browning into grapevine genomes. Such browning usually occurs as a result of the accumulation of a protein called polyphenol oxidase (PPO), which makes simple changes to the phenolic molecules known as quinones that are found in fruits, and that clump to make the browning pigment. Molecular biologists have figured out a way to turn down the production of PPO in the Sultana grape by inserting foreign DNA into its genome. Although many such attempts at genetic engineering in plants seem to be working, it remains to be seen whether genetic modification will become generally accepted as a way to produce wine grapes. Uncertainty exists because attitudes toward genetically modified plants and animals (GMOs) seem to vary by continent. People in European Union countries are wary of GMO food products, while Australians and Americans are more receptive to them. (It is interesting to contrast this with the overwhelming acceptance in Europe of the notion of evolution, while more than 50 percent of Americans reject it.) But attitudes do change. A decade ago, Australians were dead set against GMOs; now over half of those surveyed accept them. Still, current attitudes help explain why the United States and Australia are leading the way in GMO grapevines.

In 1999, a time when the genetic engineering of humans was being broadly discussed, the Princeton geneticist Lee Silver proffered an intriguing possibility. In his *Remaking Eden,* Silver suggested that unchecked genetic engineering in humans would lead to two species of humans: the gene-rich and the gene-poor. He based his Brave New World view on the understanding that only the rich would be able to take advantage of the new technology, while poor people, especially in developing countries, would not—leading to a scary Wellsian future divided between Elois and Morlocks. Although the issue with grapes is based not on wealth or availability but on opinion as to appropriateness, it is still in theory possible to envision a future dichotomy in wine production between the gene-contaminated New World and the gene-pure Old World. The wine trade has become so thoroughly globalized, however, that it is hard in practice to imagine such changes along continental lines. Future developments will clearly depend on the resilience of cultural attitudes in the face of powerful commercial imperatives.

✦ ✦ ✦

In our discussion of terroir, we saw how certain places seem to be, or to have been, regarded as particularly adapted for producing great wine. But we noted the importance of local climate in determining the perfect terroir; and the evidence indicates that climates are changing worldwide. The time scale on which the change is happening, its causes, and whether we are observing an oscillation or a long-term trend are contentious political issues. But climates *are* changing, and the vine-growing conditions at any one spot on the planet are changing right along with them.

We see evidence of this in unexpected places. The rocks that crop out in France's Champagne region and England's South Downs, on either side

View across England's South Downs, near the town of Lewes. The slopes in the middle ground, beyond the grazing sheep, may well one day be planted with vines. (Composite after a photograph by Will Harcourt-Smith)

of the English Channel, are geologically almost identical, as are the soils formed from them. Topographically the two regions are not unalike either, and geographically the difference in latitude is not much more than one degree. Yet traditionally one area produces wines that are prized the world over, while in the other sheep graze peacefully on the grassy meadows as their shepherds swill beer in the local pub.

Some twenty years ago, a French friend rather gleefully gave Ian a book titled *Les Vins de l'impossible,* which took the reader on a tour of the bizarre and unlikely places in which eccentric people somehow contrived to grow wine grapes. Among them, England took pride of place; and in 1990, when the book was published, very little wine was produced there—or even could be. Virtually everywhere, it was just too cloudy and rainy. Sunlight was inadequate, and the ripening period was too short. Even then, though, change was afoot. In the years between 1961 and 2006, the mean annual temperature in southern England increased by about 2° Celsius. That may not sound like much, but it is highly significant in terms of climate, being equivalent to a southern shift in latitude of over 300 kilometers. Largely as a result of this warming (although also facilitated by adjustments in the law), a boutique wine industry is now booming in southern England. The most successful sector of this industry produces sparkling wines. By some reckonings, the best of these sparklers are entirely comparable to their counterparts made across the Channel; occasionally an English wine will beat an illustrious marque of Champagne in a blind competition.

For now, though, the viticulturists of Champagne are not exactly unhappy. They inhabit the most northerly major winemaking region in France, at a marginal latitude for grape growing. Indeed, the tradition of making sparkling wine in Champagne probably developed because the still wines made there were a little too acidic for most palates. Since the warming trend set in conditions have improved for the Champenois, too, and the frequency of outstanding years marked for vintage (single-year) production has increased. Nonetheless, there are reasons for concern in the longer term. The two main grape varieties grown in Champagne are the Pinot Noir and the Chardonnay. These are both cool-country grapes, but they have different temperature tolerances. When the fruit is setting, Pinot Noir grows best in a narrow range of 14° to 16°C, while Chardonnay is more forgiving, doing well in temperatures up to around 18°. Currently, temperatures in Champagne remain well within the favorable range for both grapes, and immediate further warming might increase the amount of land appropriate for cultivating vines; but excessive temperature rise could at some point force a diminution of the amount of Pinot Noir grown and eventually a change in the style of the wine produced in this classic wine region.

✦ ✦ ✦

One way in which wine producers can compensate for climatic warming is by growing their vines at higher, cooler elevations. But this is not a solution available everywhere, and especially not in the gently undulating Médoc, some four and a half degrees of latitude (nearly 500 kilometers) to the south of Champagne. By some reckonings the Bordelais region as a whole is already close to the temperature maxima for the grape varieties traditionally grown there, and one alarming estimate is that the next quarter-century or so may see temperatures increase by as much as a whopping 7°C inland and 5° along the coast. Even a much more modest increase would take traditional white grapes such as the Sémillon and Sauvignon Blanc well out of their comfort zones, and eventually compel the planting of other varieties. It would probably also affect the viability of the currently grown red varietals and, at the least, have a huge effect on the style of the wines produced. At higher temperatures, and with more intense sunshine, grapes ripen more quickly; they also produce greater quantities of sugars at the expense of acids and other compounds that contribute to the structure of the wine.

Experiments in Australia have revealed that, by varying their pruning practices, growers may be able to manipulate ripening times to limit imbalances in the composition of the fruit, or to ensure that the varietals grown in a particular locality do not create logistical problems by ripening all at once. But there is a limit to the effectiveness of such interventions, and changing climates will have a significant impact on the production of wines in traditional regions. This is of special concern in areas like Bordeaux, where reputations depend on producing wines of a particular style. Wine drinkers the world over expect their clarets to have a strong tannic structure and relatively restrained fruit flavors; nobody knows how the market would react if growers in the Bordelais started to produce lush, fruit-forward wines in the style of California's hot Central Valley. The proprietors of highly reputed châteaux that cater to a clientele with specific expectations need to think about this, and many have begun to do so.

The situation in Bordeaux and elsewhere in France is complicated, however, not only by Mother Nature but also because of the country's Appellation d'Origine Contrôlée laws. These rigorously specify which varietals may be grown where, and how they may be blended. A winemaker in the Bordelais who switched varietals to accommodate to a changing environment would automatically forfeit the right to a Bordeaux designation or to any of the even more highly prized sub-denominations such as Pauillac or Saint-Estèphe. When a wine is declassified, it sells at a lower price, a reality that acts as a disincentive to vignerons to respond to climate change by growing more appropriate varietals.

The United States has its own systems of designations for winemaking areas, but since the market is organized principally by the dominant varietal, winemakers have greater flexibility in the grapes they use. Nonetheless, climate change is affecting wine production in the New World as much as in the Old. In 2006, Michael White of Utah State University and colleagues modeled future climates across the North American continent and concluded that the surface area suitable for the production of premium wines in the mainland United States could potentially decline by as much as 81 percent by the end of the century. They suggested that wine production in traditional areas would increasingly shift toward warmer-climate and lower-quality varietals, and that in many regions viticulture would be eliminated altogether because of lengthening periods of excessive heat. They also predicted that, within this century, the production of fine wines in the United States would become restricted to certain limited areas of the West Coast and the Northeast, most of which are currently handicapped by excessive rainfall.

But high temperatures and possibly associated episodes of drought and wildfire are not all that winemakers worldwide have to cope with. Along with warming comes increasing climatic instability, and if there is anything a farmer hates—and vine-growers are above all farmers—it is unpredictability. Further, vines are fussy about the conditions in which they grow, and are susceptible to disease. If conditions are unfavorable during the flowering period early in the growing season, for example, poor setting of the fruit can lead to low yields. This is not invariably bad, because as the season progresses the vines may put more effort into less fruit and produce a concentrated product. But if low fruit set is followed by anything other than ideal conditions, the results can be dire. Similarly, if it is too hot and damp during the growing season, fungal diseases may take hold, while if there is too little warmth and light as the berries develop, they may not reach the point of ripeness at which sugar levels start to increase and unpleasant organic acids decrease. In contrast, if it is too hot and moist as the grapes grow, sugar ripeness may occur before the grapes achieve phenolic ripeness, which means that the tannins and phenols in the resulting wines will remain hard and rough.

Climatic warming also increases the probability of extreme weather events, whether expressed paradoxically as winter freezing, as springtime hail, or as summer droughts. After several favorable years, the growing season of 2012 was miserable in Europe, even as record-high temperatures plagued the United States. Very dry conditions in southern Europe and extreme cold in the north played havoc with wine production. In France, production dropped by 20 percent overall, and the quality of the wines that were made varied. In Champagne, despite a string of recent successes, productivity declined by 40 percent.

Climatic modeling is a notoriously tricky process, and not all predictions agree on what lies ahead. But some trends seem to be evident, even though many vine scientists remain convinced that they will find ways of compensating for change through technological innovation. Though the time scale is hazy, in the longer term California wineries will have to consider moving their vineyards to higher elevations (although in well-established areas the best upland sites are already taken), and shifting their plantings away from cooler-clime varieties such as Riesling, Pinot Noir, and Chardonnay to those that flourish in hotter conditions, such as Nebbiolo, Zinfandel, and Carignane.

Vigorous experimentation both with the vines themselves and with the methods used to grow them may prolong the dominance of established areas, and genetic engineering techniques could also help. But even so, some think that within a few decades the area of the Napa Valley suited to fine-wine production will have declined by as much as 50 percent. At the same time, some of the cooler parts of Oregon (notably the Willamette Valley) and Washington State (Walla Walla, east of the Cascades, where some fine Cabernet Sauvignons are already made, and even the unlikely area around Puget Sound near Seattle, where vines would barely grow a mere half-century ago) will have moved to the fore in West Coast wine production. Even the Okanagan Valley of Canada's British Columbia may go from being a marginal to a prominent producer. In the eastern United States, the Finger Lakes, the lower Hudson Valley, and Long Island are all poised to move up in reputation as climates warm.

Worldwide, a similar shift from traditionally famous wine-producing areas to currently obscure regions is also predicted. Tasmania and parts of the South Island of New Zealand are expected to assume greater importance in Australasian fine wine production, while in Europe we have seen that southern England is projected to become more significant (if vineyards can compete with other forms of land use). In hot dry regions like Portugal and southern Spain, wine production has already begun the move to higher elevations. Altogether we are in a period of extraordinary flux, in which wine producers are going to have to be nimble if they wish to cope with potential major changes that will include an increased frequency of disasters such as wildfire and flooding.

✦ ✦ ✦

Does this mean that we consumers will need to learn to enjoy wines of different styles, made from different grapes, from the ones we are used to? If present climatic trends continue, in the long term the answer is almost certainly a qualified yes. But nobody knows how far away the long term might be, and we cannot predict what human ingenuity might devise to mitigate the situation. A best guess might be that wine producers, who have an enormous incentive to stay where they are, will use every means at their disposal to provide a stable product for wine consumers, who tend to know what they like (or at least think they do). But adaptation to a changing climate will involve a lot of work. Gregory Jones, a climatologist at the University of Oregon who has estimated that in the Northern Hemisphere the broad geographical swaths of territory suitable for winemaking will move northward by between 275 and 550 kilometers over the next hundred years, has pointed the way ahead: "It will be those . . . that are the most aware, that experiment with both methods and technology—in plant breeding and genetics, in the field, and in processing—that will have the greatest latitude of adaptation."

So just as the march of technology has begun to offer winemakers an infinity of possibilities, it seems that viticulturists will find themselves in much the same position as the Red Queen in *Through the Looking Glass,* whose subjects had to run as fast as they could to stay in the same place. And before long, it seems, many winegrowers will similarly find themselves obliged to change their vineyard and production procedures in order to keep their wines looking and tasting the same. Oenophiles, by and large a pretty conservative group in their vinous tastes and expectations, will be hoping they succeed.

FRANKEN-VINES AND CLIMATE CHANGE

Annotated Bibliography

There is a huge literature on wine. Below we provide a chapter-by chapter annotated listing of the major sources consulted and quoted in the writing of this book.

CHAPTER 1. VINOUS ROOTS

The best available overviews of ancient wine and other fermented beverages are McGovern, *Ancient Wine* and *Uncorking the Past*. The evidence for early wine production at Areni is given in Barnard et al., "Chemical Evidence." The Xenophon quotation is from book 4. For general information on Abu Hureya see Moore, Hillman, and Legge, *Village on the Euphrates*; for Hajji Firuz Tepe wine residues consult McGovern, Glusker, and Exner, "Neolithic Resinated Wine." Vouillamoz et al., "Genetic Characterization," looks at traditional Caucasian cultivars. For an overview of wine in ancient Egypt, see Poo, *Wine and Wine Offering*, and for an analysis of ancient Egyptian herbal wines see McGovern, Mirzoian, and Hall, "Ancient Egyptian Herbal Wines." For early evidence of viticulture, see Jiang et al., "Evidence for Early Viticulture in China," and McGovern et al., "Beginning of Viniculture in France." Standage, *History of the World in Six Glasses*, provides an entertaining general survey of wine consumption in the classical world, and Unwin, *Wine and the Vine*, and Phillips, *Short History of Wine*, provide more detail. The Franklin quotation is from a letter written to the abbé André Morellet in 1787. A comprehensive review of Prohibition worldwide is furnished by Blocker, Fahey, and Tyrrell, *Alcohol and Temperance in Modern History*.

Barnard, H., A. N. Dooley, G. Areshian, B. Gasparyan, and K. F. Faul."Chemical Evidence for Wine Production Around 4000 BCE in the Late Chalcolithic Near Eastern Highlands." *Journal of Archaeological Science* 38 (2011): 977–984.

Blocker, Jack. S., Jr., David M. Fahey, and Ian R. Tyrrell, eds. *Alcohol and Temperance in Modern History: An International Encyclopedia.* Santa Barbara, Calif.: ABC-CLIO, 2003.

Jiang, H.-E., Y.-B. Zhang, X. Li, Y.-F. Yao, et al. "Evidence for Early Viticulture in China: Proof of a Grapevine (*Vitis vinifera* L., Vitaceae) in the Yanghai Tombs, Xinjiang." *Journal of Archaeological Science* 36 (2009): 1458–1465.

McGovern, Patrick E. *Ancient Wine: The Search for the Origins of Viticulture.* Princeton: Princeton University Press, 2003.

McGovern, Patrick E. *Uncorking the Past: The Quest for Wine, Beer and Other Alcoholic Beverages.* Berkeley: University of California Press, 2009.

McGovern, P. E., D. L. Glusker, and L. J. Exner. "Neolithic Resinated Wine." *Nature* 381 (1996): 480–481.

McGovern, P. E., B. P. Luley, N. Rovira, A. Mirzoian, et al. "Beginning of Viniculture in France." *Proceedings of the National Academy of Sciences of the United States of America* 110, no. 25 (2013): 10147–10152.

McGovern, P. E., A. Mirzoian, and G. R. Hall. "Ancient Egyptian Herbal Wines." *Proceedings of the National Academy of Sciences of the United States of America* 106 (2009): 7361–7366.

Moore, A. M. T., G. C. Hillman, and A. J. Legge. *Village on the Euphrates: From Foraging to Farming at Abu Hureyra.* Oxford: Oxford University Press, 2000.

Phillips, Rod. *A Short History of Wine.* London: Allen Lane 2000.

Poo, Mu-chou. *Wine and Wine Offering in the Religion of Ancient Egypt.* London: Kegan Paul, 1995.

Standage, Tom. *A History of the World in Six Glasses.* New York: Walker, 2005.

Unwin, Tim. *Wine and the Vine: An Historical Geography of Viticulture and the Wine Trade.* London: Routledge, 1996.

Vouillamoz, J. F., P. E. McGovern, A. Ergul, G. Söylemezoğlu, et al. "Genetic Characterization and Relationships of Traditional Grape Cultivars from Transcaucasia and Anatolia." *Plant Genetic Resources* 4, no. 2 (2006): 144–158.

Xenophon. *Anabasis: The March Up Country.* Trans. H. G. Dakyns. El Paso: El Paso Norte Press, 2007.

CHAPTER 2. WHY WE DRINK WINE

For added longevity in fruit fly "drinkers," see Starmer, Heed, and Rockwood-Sluss, "Extension of Longevity"; for self-medication see Milan, Kacsoh, and Schlenke, "Alcohol Consumption as Self-medication"; and for sexual deprivation and ethanol preference see Shohat-Ophir et al., "Sexual Deprivation Increases Ethanol Intake." The tippling habits of tree shrews were reported in Wiens et al., "Chronic Intake of Fermented Floral Nectar." For general discussions of ethanol consumption and alcoholism see Levey, "Evolutionary Ecology of Ethanol Production"; Dudley, "Ethanol, Fruit Ripening, and the Historical Origins of Human Alcoholism"; and Milton, "Ferment in the Family Tree." For ethanol and foraging see Dominy, "Fruits, Fingers and Fermentation." For specific statements of the "drunken monkey hypothesis," consult Dudley, "Evolutionary Origins of Human Alcoholism," and Stephens and Dudley, "Drunken Monkey Hypothesis." For enzyme change in common ancestor of African apes and humans see Carrigan et al., "Hominids."

Carrigan, M. A., Uryasev, O., Frye, C. B., Eckman, B. L., et al. "Hominids Adapted to Metabolize Ethanol Long Before Human-directed Fermentation." *Proceedings of the National Academy of Sciences of the United States of America* 112, no. 2 (2014): 458–463.

Dominy, N. J. "Fruits, Fingers and Fermentation: The Sensory Cues Available to Foraging Primates." *Integrative and Comparative Biology* 44 (2004): 295–303.

Dudley, R. "Ethanol, Fruit Ripening, and the Historical Origins of Human Alcoholism in Primate Frugivory." *Integrative and Comparative Biology* 44 (2004): 315–323.

Dudley, R. "Evolutionary Origins of Human Alcoholism in Primate Frugivory." *Quarterly Review of Biology* 75 (2000): 3–15.

Levey, D. J. "The Evolutionary Ecology of Ethanol Production and Alcoholism." *Integrative and Comparative Biology* 44 (2004): 284–289.

Milan, N. F., B. Z. Kacsoh, and T. A. Schlenke. "Alcohol Consumption as Self-medication Against Blood-borne Parasites in the Fruit Fly." *Current Biology* 22 (2012): 488–493.

Milton, K. "Ferment in the Family Tree: Does a Frugivorous Dietary Heritage Influence Contemporary Patterns of Human Ethanol Use?" *Integrative and Comparative Biology* 44 (2004): 304–314.

Shohat-Ophir, G., K. R. Kaun, R. Azanchi, and U. Heberlein. "Sexual Deprivation Increases Ethanol Intake in *Drosophila*." *Science* 335 (2012): 1351–1355.

Starmer, W. T., W. B. Heed, and E. S. Rockwood-Sluss. "Extension of Longevity in *Drosophila mojavensis* by Environmental Ethanol: Differences Between Subraces." *Proceedings of the National Academy of Sciences of the United States of America* 74 (1977): 387–391.

Stephens, D., and R. Dudley. "The Drunken Monkey Hypothesis: The Study of Fruit-eating Animals Could Lead to an Evolutionary Understanding of Human Alcohol Abuse." *Natural History* 113 (2004): 40–44.

Wiens, F., A. Zitzmann, M.-A. Lachance, M. Yegles, et al., "Chronic Intake of Fermented Floral Nectar by Wild Treeshrews." *Proceedings of the National Academy of Sciences of the United States of America* 105 (2008): 10426–10431.

CHAPTER 3. WINE IS STARDUST

This chapter is based on some basic biology and biochemistry that can be obtained in any good high school biology textbook. We self-servingly offer as an example DeSalle and Heithaus, *Biology*. More specific treatment of the chemistry, biochemistry, and biology involved can be found in two superb books by Nick Lane, *Oxygen* and *Life Ascending* (which has a wonderful description of how photosynthesis works), and, on wine specifically, in Margalit, *Concepts in Wine Chemistry*. Zuckerman's discovery of extraterrestrial alcohol molecules is amusingly discussed in Tyson, "Milky Way Bar," which is also the source of the quotation. The biology of convergence in general is presented nicely in Huston, *Biological Diversity*, while the biology and genetics of gene expression in grape material can be found in Grimplet et al., "Tissue-specific mRNA Expression Profiling." The phylogenetics of plants discussed in this chapter can be found in Lee et al., "Functional Phylogenomics View of the Seed Plants."

DeSalle, Rob, and Michael R. Heithaus. *Biology*. New York: Holt, Rinehart and Winston, 2007.

Felger, R., and J. Henrickson. "Convergent Adaptive Morphology of a Sonoran Desert Cactus (*Peniocereus striatus*) and an African Spurge (*Euphorbia cryptospinosa*)." *Haseltonia* 5 (1977): 77–85.

Grimplet, J., L. G. Deluc, R. L. Tillett, M. D. Wheatley, et al. "Tissue-specific mRNA Expression Profiling in Grape Berry Tissues." *BMC Genomics* 8, no. 1 (2007): 187.

Huston, Michael A., *Biological Diversity: The Coexistence of Species*. Cambridge: Cambridge University Press, 1994.

Lane, Nick. *Life Ascending: The Ten Great Inventions of Evolution*. London: Profile, 2010.

Lane, Nick. *Oxygen: The Molecule That Made the World*. Oxford: Oxford University Press, 2002.

Lee, E. K., A. Cibrian-Jaramillo, S. O. Kolokotronis, M. S. Katari, et al. "A Functional Phylogenomics View of the Seed Plants." *PLoS Genet* 7, no. 12 (2011):e1002411.

Margalit, Yair. *Concepts in Wine Chemistry*. 3rd ed. San Francisco: Wine Appreciation Guild, 2012.

Tyson, N. D. "The Milky Way Bar." *Natural History* 103 (August 1995): 16–18.

CHAPTER 4. GRAPES AND GRAPEVINES

The *Revisio* is Kuntze, *Revisio generum plantarum vascularium*. The biology of seeds and the discovery of seedlessness is discussed in a recent publication by Lora et al., "Seedless Fruits." Darwin's "Abominable Mystery" is described from a historical perspective in Friedman, "Meaning of Darwin's 'Abominable Mystery,'" and living fossils are adeptly discussed in Fortey, *Survivors*. There is a huge literature on the origins of grapes and their relationships using molecular techniques. Key references include This, Lacombe, and Thomas, "Historical Origins"; Soejima and Wen, "Phylogenetic Analysis"; Tröndle et al., "Molecular Phylogeny"; Zecca et al., "Timing and Mode of Evolution"; Myles et al., "Genetic Structure"; Le Cunff et al., "Construction of Nested Genetic Core Collections"; Bacilieri et al., "Genetic Structure" (for the findings of Laucou's team); de Andrés et al., "Molecular Characterization of Grapevine Rootstocks" (for the findings of Zapater's team); Arroyo-García et al., "Multiple Origins"; and Terral et al., "Evolution and History."

Arroyo-García, R., L. Ruiz-García, L. Bolling, R. Ocete, et al. "Multiple Origins of Cultivated Grapevine (Vitis vinifera L. ssp. sativa) Based on Chloroplast DNA Polymorphisms." *Molecular Ecology* 15, no. 12 (2006): 3707–3714.

Bacilieri, R., T. Lacombe, L. Le Cunff, M. Di Vecchi-Staraz, et al. "Genetic Structure in Cultivated Grapevines Is Linked to Geography and Human Selection." *BMC Plant Biology* 13, no. 1 (2013): 25.

de Andrés, M. T., J. A. Cabezas, M. T. Cervera, J. Borrego, et al. "Molecular Characterization of Grapevine Rootstocks Maintained in Germplasm Collections." *American Journal of Enology and Viticulture* 58, no. 1 (2007): 75–86.

Fortey, Richard. *Survivors: The Animals and Plants That Time Has Left Behind*. London: Harper Collins, 2011.

Friedman, W. E. "The Meaning of Darwin's 'Abominable Mystery.'" *American Journal of Botany* 96, no. 1 (2009): 5–21.

Kuntze, Otto. *Revisio generum plantarum vascularium omnium atque cellularium multarum secundum Leges nomenclaturae internationales cum enumeratione plantarum exoticarum in itinere mundi collectarum: Pars I-[III]*. Vol. 3A. Leipzig: Felix, 1893.

Le Cunff, L., A. Fournier-Level, V. Laucou, S. Vezzulli, et al. "Construction of Nested Genetic Core Collections to Optimize the Exploitation of Natural Diversity in Vitis vinifera L. subsp. sativa." *BMC Plant Biology* 8, no. 1 (2008): 31.

Lora, J., J. I. Hormaza, M. Herrero, and C. S. Gasser. "Seedless Fruits and the Disruption of a

Conserved Genetic Pathway in Angiosperm Ovule Development." *Proceedings of the National Academy of Sciences of the United States of America* 108, no. 13 (2011): 5461–5465.

Myles, S., A. R. Boyko, C. L. Owens, P. J. Brown, et al. "Genetic Structure and Domestication History of the Grape." *Proceedings of the National Academy of Sciences of the United States of America* 108, no. 9 (2011): 3530–3535.

Soejima, A., and J. Wen. "Phylogenetic Analysis of the Grape Family (Vitaceae) Based on Three Chloroplast Markers." *American Journal of Botany* 93, no. 2 (2006): 278–287.

Terral, J.-F., E. Tabard, L. Bouby, S. Ivorra, et al. "Evolution and History of Grapevine (*Vitis vinifera*) Under Domestication: New Morphometric Perspectives to Understand Seed Domestication Syndrome and Reveal Origins of Ancient European Cultivars." *Annals of Botany* 105, no. 3 (2010): 443–455.

This, P., T. Lacombe, and M. R. Thomas. "Historical Origins and Genetic Diversity of Wine Grapes." *Trends in Genetics* 22, no. 9 (2006): 511–519.

Trias-Blasi, A., J. A. N. Parnell, and T. R. Hodkinson. "Multi-gene Region Phylogenetic Analysis of the Grape Family (Vitaceae)." *Systematic Botany* 37, no. 4 (2012): 941–950.

Tröndle, D., S. Schröder, H.-H. Kassemeyer, C. Kiefer, et al. "Molecular Phylogeny of the Genus Vitis (Vitaceae) Based on Plastid Markers." *American Journal of Botany* 97, no. 7 (2010): 1168–1178.

Zecca, G., J. R. Abbott, W.-B. Sun, A. Spada, et al. "The Timing and the Mode of Evolution of Wild Grapes (Vitis)." *Molecular Phylogenetics and Evolution* 62, no. 2 (2012): 736–747.

CHAPTER 5. YEASTY FEASTS

The dynamics of yeast and fungal systematics can be found in James et al., "Reconstructing the Early Evolution of Fungi" (for the Vilgalys team); Liti et al., "Population Genomics"; and Stefanini et al., "Role of Social Wasps" (for the Cavalieri team). See the last of these for the role of hornets. The quotation on microbial cities comes from Tiedje, "20 Years Since Dunedin."

James, T. Y., F. Kauff, C. L. Schoch, P. B. Matheny, et al. "Reconstructing the Early Evolution of Fungi Using a Six-gene Phylogeny." *Nature* 443, no. 7113 (2006): 818–822.

Liti, G., D. M. Carter, A. M. Moses, J. Warringer, et al. "Population Genomics of Domestic and Wild Yeasts." *Nature* 458, no. 7236 (2009): 337–341.

Stefanini, I., L. Dapporto, J.-L. Legras, A. Calabretta, et al. "Role of Social Wasps in *Saccharomyces cerevisiae* Ecology and Evolution." *Proceedings of the National Academy of Sciences of the United States of America* 109, no. 33 (2012): 13398–13403.

Tiedje, J. M. "20 Years Since Dunedin: The Past and Future of Microbial Ecology." In *Microbial Biosystems: New Frontiers. Proceedings of the 8th International Symposium on Microbial Ecology,* ed. C. R. Bell, M. Brylinsky, and P. Johnson-Green. Halifax: Atlantic Canada Society for Microbial Ecology, 1999. Available at http://plato.acadiau.ca/isme/Symposium29/tiedje .PDF.

CHAPTER 6. INTERACTIONS

The historical writings referenced in this chapter include Theophrastus, *Enquiry into Plants;* Darwin, *On the Various Contrivances;* and Wainwright and Lederberg, "History of Microbiology" (which contains material on Martinus Beijerinck and Sergei Winogradsky). Microbial components of wine and the genetics of wine color are discussed in Moter and Göbel, "Fluorescence in Situ Hybridization"; Renouf, Claisse, and Lonvaud-Funel, "Inventory and Monitoring"; Barata, Malfeito-Ferreira, and Loureiro, "Microbial Ecology" (the "researchers in Portugal"); Shimazaki et al., "Pink-colored Grape Berry"; and Bokulich et al., "Microbial Biogeography."

Barata, A., M. Malfeito-Ferreira, and V. Loureiro. "The Microbial Ecology of Wine Grape Berries." *International Journal of Food Microbiology* 153, no. 2 (2012): 243–259.

Bokulich, N. A., J. H. Thorngate, P. M. Richardson, and D. A. Mills. "Microbial Biogeography of Wine Grapes Is Conditioned by Cultivar, Vintage, and Climate." *Proceedings of the National Academy of Sciences of the United States of America* 111, no. 1 (2014): E139–E148.

Darwin, Charles. *On the Various Contrivances by Which British and Foreign Orchids Are Fertilised by Insects: And on the Good Effects of Intercrossing.* London: Murray, 1862.

Moter, A., and U. B. Göbel. "Fluorescence in Situ Hybridization (FISH) for Direct Visualization of Microorganisms." *Journal of Microbiological Methods* 41, no. 2 (2000): 85–112.

Renouf, V., O. Claisse, and A. Lonvaud-Funel. "Inventory and Monitoring of Wine Microbial Consortia." *Applied Microbiological Biotechnology* 75, no. 1 (2007): 149–164.

Shimazaki, M., K. Fujita, H. Kobayashi, and S. Suzuki. "Pink-colored Grape Berry Is the Result of Short Insertion in Intron of Color Regulatory Gene." *PLoS One* 6, no. 6 (2011): e21308.

Theophrastus. *Enquiry into Plants, Books 1–5.* Trans. A. F. Hort. Cambridge: Harvard University Press, 1916.

Theophrastus. *Enquiry into Plants, Books 6–9; Treatise on Odours; Concerning Weather Signs.* Trans. A. F. Hort. Cambridge: Harvard University Press, 1916.

Wainwright, Milton, and Joshua Lederberg. "History of Microbiology." *Encyclopedia of Microbiology*, 2:419–437. New York: Academic Press, 1992.

CHAPTER 7. THE AMERICAN DISEASE

A classic of phylloxera literature is Planchon's *Vignes américaines.* There are several excellent current books on phylloxera, among them those by Campbell (*Botanist and the Vintner*) and Gale (*Dying on the Vine*). The latter, in particular, contains a large bibliography pointing to the extensive specialized literature on the subject. Important work on the life cycle of the phylloxera insect was reported in Granett, Bisabri-Ershadi, and Carey, "Life Tables of Phylloxera." The Prial quote is from his "After Phylloxera," and a radical appraisal of the long-term health effects of the French phylloxera outbreak was recently published by Banerjee et al. ("Long-run Health Impacts").

Banerjee, A., E. Duflo, G. Postel-Vinay, and T. Watts. "Long-run Health Impacts of Income Shocks: Wine and Phylloxera in Nineteenth-century France." *Review of Economics and Statistics* 92 (2013): 714–728.

Campbell, Christy. *The Botanist and the Vintner: How Wine Was Saved for the World*. New York: Algonquin Books of Chapel Hill, 2004.

Gale, George D., Jr. *Dying on the Vine: How Phylloxera Transformed Wine*. Berkeley: University of California Press, 2011.

Granett, J., B. Bisabri-Ershadi, and J. Carey. "Life Tables of Phylloxera on Resistant and Susceptible Rootstocks." *Entomology Experimental and Applied* 34, no. 1 (1983): 13–19.

Planchon, Jules-Émile. *Les Vignes américaines: leur culture, leur résistance au Phylloxéra et leur avenir en Europe*. 1875. Available at amazon.com in several facsimile reprints.

Prial, F. "After Phylloxera, the First Taste of a Better Grape." *New York Times*, May 5, 1999.

CHAPTER 8. THE REIGN OF TERROIR

A socioeconomic consideration of the concept of terroir and the practical challenges it imposes is provided in Barham, "Translating Terroir." A good general discussion is found in Sommers, *Geography of Wine*. An important overview of terroir in the French wine lands, including Champagne, the Bordelais, and Burgundy, is Wilson, *Terroir*, and classic works on Cahors and the soils of Bordeaux are Baudel, *Vin de Cahors*, and Seguin, *Influence des facteurs naturels*, respectively. A good technical treatment of soils is White, *Soils for Fine Wines*, and a splendidly accessible treatment of terroir in the Napa Valley is Swinchatt and Howell, *Winemaker's Dance*.

Barham, E. "Translating Terroir: The Global Challenge of French AOC Labeling." *Journal of Rural Studies* 19 (2003): 127–138.

Baudel, José. *Le Vin de Cahors*. Luzech: Cotes d'Olt, 1972.

Seguin, Gérard. *Influence des facteurs naturels sur les caractères des vins*. Paris: Dunod, 1971.

Sommers, Brian J. *The Geography of Wine: How Landscapes, Cultures, Terroir and the Weather Make a Good Drop*. New York: Plume, 2008.

Swinchatt, Jonathan, and David G. Howell, *The Winemaker's Dance: Exploring Terroir in the Napa Valley*. Berkeley: University of California Press, 2004.

White, Robert E. *Soils for Fine Wines*. Oxford: Oxford University Press, 2003.

Wilson, James E. *Terroir: The Role of Geology, Climate, and Culture in the Making of French Wines*. Berkeley: University of California Press, 1998.

CHAPTER 9. WINE AND THE SENSES

Excellent overall references for how wine piques our senses can be found in McGovern, *Uncorking the Past*, and Shepherd, *Neurogastronomy*. Historical references for this chapter include Piccolino and Wade, "Galileo Galilei's Vision of the Senses"; Liger-Belair, *Uncorked*; McCoy, *Emperor of Wine*; and Lukacs, *Inventing Wine*. Yokoyama, "Molecular Evolution," provides an excellent review of color vision in vertebrates, while Turin and Yoshii, "Structure-odor Relations," details the process of odor perception, and tetrachromacy in humans was reported by Nagy et al. (1981). Peynaud's classic work on the sensory evaluation of wine, *The Taste of Wine*, is well worth reading if you can find it. For the relationship between the shape of the glass and Champagne bubbles, see Liger-Belair, *Uncorked*. On the influence of difficult-to-pronounce winery names see Mantonakis and Galiffi, "Does How Fluent a Winery Name Sounds Affect Taste Perception?" For a lively discussion of the Robert Parker phenomenon, see Lukacs, *Inventing Wine*. A good general reference on neuroeconomics is Glimcher, *Foundations of Neuroeconomic Analysis*. References for the specific neuroeconomic studies discussed in this chapter can be found in Plassman et al., "Marketing Actions"; Mantonakis et al., "False Beliefs Can Shape Current Consumption"; DeMello and Pires Gonçalves, "Message on a Bottle"; Mantonakis and Galiffi, "Does How Fluent a Winery Name Sounds Affect Taste Perception?"; and Almenberg and Almenberg, "Appendix 2" (for the Swedish-Yale experiment and quotation).

Almenberg, Johan, and Anna Dreber Almenberg, "Appendix 2: Experimental Conclusions." In *The Wine Trials: A Fearless Critic Book*, ed. Robin Goldstein, with Alexis Herschkowitsch. Austin, Tex.: Fearless Critic Media, 2008.

De Mello, L., and R. Pires Gonçalves. "Message on a Bottle: Colours and Shapes in Wine Labels." *Munich Personal RePEc Archive*, Paper No. 13122 (2009).

Glimcher, Paul W. *Foundations of Neuroeconomic Analysis*. Oxford: Oxford University Press, 2011.

Liger-Belair, Gérard. *Uncorked: The Science of Champagne*. Rev. ed. Princeton: Princeton University Press, 2013.

Lukacs, Paul. *Inventing Wine: A New History of One of the World's Most Ancient Pleasures*. New York: Norton, 2012.

Mantonakis, A., and B. Galiffi. "Does How Fluent a Winery Name Sounds Affect Taste Perception?" *Sixth AWBR International Conference Abstracts* (2011): 1–7.

Mantonakis, A., A. Wudarzewski, D. M. Bernstein, S. L. Clifasefi, and E. F. Loftus. "False Beliefs Can Shape Current Consumption." *Psychology* 4, no. 3 (2013): 302.

McCoy, Elin. *The Emperor of Wine: The Rise of Robert M. Parker, Jr., and the Reign of American Taste*. New York: Ecco, 2005.

McGovern, Patrick E. *Uncorking the Past: The Quest for Wine, Beer, and Other Alcoholic Beverages*. Berkeley: University of California Press, 2009.

Nagy, A. L., D. I .A. MacLeod, N. E Heyneman, and A. Eisner. "Four Cone Pigments in Women Heterozygous for Color Deficiency." *Journal of the Optical Society of America* 71 (1981): 719–722.

Peynaud, Émile. *The Taste of Wine: The Art and Science of Wine Appreciation*. San Francisco: Wine Appreciation Guild, 1997.

Piccolino, M., and N. J. Wade. "Galileo Galilei's Vision of the Senses." *Trends in Neuroscience* 31, no. 11 (2008): 585–590.

Plassmann, H., J. O'Doherty, B. Shiv, and A. Rangel. "Marketing Actions Can Modulate Neural Representations of Experienced Pleasantness." *Proceedings of the National Academy of Sciences of the United States of America* 105, no. 3 (2008): 1050–1054.

Shepherd, Gordon M. *Neurogastronomy: How the Brain Creates Flavor and Why It Matters*. New York: Columbia University Press, 2012.

Turin, Luca, and Fumiko Yoshii. "Structure-odor Relations: A Modern Perspective." In *Handbook of Olfaction and Gustation*, 275–294. Hoboken, N.J.: Wiley-Blackwell, 2003.

Yokoyama, S. "Molecular Evolution of Color Vision in Vertebrates." *Gene* 300, no. 1 (2002): 69–78.

CHAPTER 10. VOLUNTARY MADNESS

Jen Kirkman's performance can be found at the Funny or Die website (http://www.funnyordie
.com/videos/d47e6a33a5/drunk-history-vol-5-w-will-ferrell-don-cheadle-zooey-deschanel).
If the room is spinning and you need advice, the following archive website might be helpful:
http://arstechnica.com/civis/viewtopic.php?f=23&t=306174. The biology of the liver on alcohol is discussed at length in Epstein, "Alcohol's Impact." The genetics of alcohol processing and alcoholism in humans are discussed in Lu and Cederbaum, "CYP2E1 and Oxidative Liver Injury"; Oota et al., "Evolution and Population Genetics of the ALDH2 Locus"; Mulligan et al., "Allelic Variation"; and Bierut et al., "Genome-wide Association Study of Alcohol Dependence." See also Francis Crick's wonderful treatise on neurobiology, *Astonishing Hypothesis*.

Bierut, L. J., A. Agrawal, K. K. Bucholz, K. F. Doheny, et al. "A Genome-wide Association Study of Alcohol Dependence." *Proceedings of the National Academy of Sciences of the United States Of America* 107, no. 11 (2010): 5082–5087.

Crick, Francis. *Astonishing Hypothesis: The Scientific Search for the Soul*. New York: Scribner's, 1995.

Epstein, M. "Alcohol's Impact on Kidney Function." *Alcohol Health Research World* 21 (1997): 84–92.

Hinrichs, A. L., J. C. Wang, B. Bufe, J. M. Kwon, et al. "Functional Variant in a Bitter-taste Receptor (hTAS2R16) Influences Risk of Alcohol Dependence." *American Journal of Human Genetics* 78, no. 1 (2006): 103–111.

Lu, Y., and A. I. Cederbaum. "CYP2E1 and Oxidative Liver Injury by Alcohol." *Free Radicals in Biology and Medicine* 44, no. 5 (2008): 723–738.

Mulligan, C., R. W. Robin, M. V. Osier, N Sambuughin, et al. "Allelic Variation at Alcohol Metabolism Genes (ADH1B, ADH1C, ALDH2) and Alcohol Dependence in an American Indian Population." *Human Genetics* 113, no. 4 (2003): 325–336.

Oota, H., A. J. Pakstis, B. Bonne-Tamir, D. Goldman, et al. "The Evolution and Population Genetics of the ALDH2 Locus: Random Genetic Drift, Selection, and Low Levels of Recombination." *Annals of Human Genetics* 68, no. 2 (2004): 93–109.

CHAPTER 11. BRAVE NEW WORLD

There is a large, expanding literature on the technology of winemaking (although much information on this subject is proprietary). Excellent though technical available works are Winkler et al., *General Viticulture;* Jackisch, *Modern Winemaking;* and Margalit, *Winery Technology,* while a very accessible work is Cox, *From Vines to Wines.* Peynaud's classic work (Spencer and Peynaud, *Knowing and Making Wine*) still remains a mandatory read, and an

American classic is Amerine's *Technology of Wine Making*. A valuable work that makes reference to many current techniques is Bird, *Understanding Wine Technology*, and an excellent and accessible overview is Goode, *Science of Wine*. Benjamin Wallace's instant classic on wine fakery, *The Billionaire's Vinegar*, is an engaging account of the kind of skullduggery that the transformation of wines into valuable collectibles has encouraged, and the perusal of almost any issue of *Wine Spectator* will yield yet more examples.

Amerine, Maynard A. *The Technology of Wine Making*. 4th ed. Westport, Conn.: Avi, 1980.

Bird, David. *Understanding Wine Technology: The Science of Wine Explained*. 3rd ed. San Francisco: Wine Appreciation Guild, 2011.

Cox, Jeff. *From Vines to Wines: The Complete Guide to Growing Grapes and Making Your Own Wine*. 5th ed. North Adams, Mass.: Storey, 2015.

Goode, Jamie. *The Science of Wine: From Vine to Glass*. 2nd ed. Berkeley: University of California Press, 2014.

Jackisch, Philip. *Modern Winemaking*. Ithaca: Cornell University Press, 1985.

Margalit, Yair. *Winery Technology and Operations: A Handbook for Small Wineries*. San Francisco: Wine Appreciation Guild, 1996.

Spencer, Alan F., and Émile Peynaud. *Knowing and Making Wine*. New York: Houghton Mifflin Harcourt, 1984.

Wallace, Benjamin. *The Billionaire's Vinegar: The Mystery of the World's Most Expensive Bottle of Wine*. New York: Crown, 2008.

Winkler, A. J., James A. Cook, W. M. Kliewer, and Lloyd A. Lider. *General Viticulture*. Rev. ed. Berkeley: University of California Press, 1974.

CHAPTER 12. FRANKEN-VINES AND CLIMATE CHANGE

The basics of the grape genome are explained by Jaillon et al., "Grapevine Genome Sequence." See Reustle and Büchholz, "Recent Trends," for an overview of GMO grapes. Mulwa et al., "Agrobacterium-mediated Transformation," discusses the case of the modified Chancellor grape. European attitudes to GMO organisms are discussed in Pardo, Midden, and Miller, "Attitudes Toward Biotechnology." The literature on wine and climate change is becoming larger daily. Gregory Jones and colleagues have written extensively on the potential impact in the United States (Jones, "Climate Change in the Western United States") and worldwide (Jones et al., "Climate Change and Global Wine Quality," from which the quotation comes); Webb, Whetton, and Barlow ("Modelled Impact") have sounded the alarm for Australia. Hayhoe et al. have bracketed alarming predicted effects on California in "Emissions Pathways," and White et al. make some pretty dire forecasts for premium wine production throughout the United States for the coming century in "Extreme Heat." Hannah et al., "Climate Change," warns of the need to modify viticultural practices in the face of climatic warming. Goode ("Fruity with a Hint of Drought") surveys the complexities of the situation in an accessible way.

Goode, J. "Fruity with a Hint of Drought." *Nature* 492 (2012): 351–353.

Hannah, L., P. R. Roehrdanz, M. Ikegami, A. V. Shepard, et al. "Climate Change, Wine, and Conservation." *Proceedings of the National Academy of Sciences of the United States of America* 110, no. 17 (2013): 6907–6912.

Hayhoe, K., D. Cayan, C. B. Field, P. C. Frumhoff, et al. "Emissions Pathways, Climate Change, and Impacts on California." *Proceedings of the National Academy of Sciences of the United States of America* 101 (2004): 12422–12427.

Huetz de Lamps, Alain. *Les Vins de l'impossible.* Grenoble: Glénat, 1990.

Jaillon, O., J.-M. Aury, B. Noel, A. Policriti, et al., for the French-Italian Public Consortium for Grapevine Genome Characterization. "The Grapevine Genome Sequence Suggests Ancestral Hexaploidization in Major Angiosperm Phyla." *Nature* 449, no. 7161 (2007): 463–467.

Jones, G. V. "Climate Change in the Western United States Grape Growing Regions." *Acta Horticultura* 689 (2005): 41–59.

Jones, G. V., M. A. White, O. R. Cooper, and K. Storchmann. "Climate Change and Global Wine Quality." *Climatic Change* 73 (2005): 319–343.

Mulwa, R. M. S., M. A. Norton, S. K. Farrand, and R. M. Skirvin. "Agrobacterium-mediated Transformation and Regeneration of Transgenic Chancellor's Wine Grape Plants Expressing the tfdA Gene." *Vitis-Geilweilerhof* 46, no. 3 (2007): 110.

Pardo, R., C. Midden, and J. D. Miller. "Attitudes Toward Biotechnology in the European Union." *Journal of Biotechnology* 98, no. 1 (2002): 9–24.

Reustle, G. M., and G. Büchholz. "Recent Trends in Grapevine Genetic Engineering." In *Grapevine Molecular Physiology and Biotechnology,* ed. Kalliopi A. Roubelakis-Angelakis, 495–508. Amsterdam: Springer Netherlands, 2009.

Silver, Lee M. *Remaking Eden.* New York: Avon, 1998.

Webb, L. B., P. H. Whetton, and E. W. R. Barlow. "Modelled Impact of Future Climate Change on the Phenology of Winegrapes in Australia." *Australian Journal of Grape and Wine Research* 13 (2007): 165–175.

White, M. A., N. S. Diffenbaugh, G. V. Jones, J. S. Pal, and F. Giorgi. "Extreme Heat Reduces and Shifts Unites States Premium Wine Production in the 21st Century." *Proceedings of the National Academy of Sciences of the United States of America* 103 (2006): 11217–11222.

Index

Page numbers in *italics* refer to illustrations.

South Island, New Zealand, 224–25

Soviet Union, 4

Spain, 16, 122, 129, 151, 152, 225

sparkling wines, 54, 156, 168–69, 197, 221.
    See also Champagne

spectrophotometer, 160–61

sperm, 69, 72

spice, 166

spoilage, 197

spores, 73

Sporidiales, 95

Sporidiobolaceae, 95

stainless steel, 54, 199, 202

starches, 9, 44–45

stele, of grape, 64

stellate cells, 184

stem cells, 69

Step Pyramid of Djoser, 12

stick notation, 39

Stilton cheese, 166

stinkhorns, 94

stomach, 182–83

Streptococcus thermophilus, 51

structural genes, 106

STRUCTURE analysis, 96

sturgeon, 213

stylar remnant, 64

subspecies, 76

sucrose, 42, 45, 65

sugars, 9, 12, 35–36, 51–52, 54, 63, 72, 93, 99,
    115, 162, 184; bacterial effects on, 49;
    dosage with, 55, 197; ethanol and, 25,
    29, 42, 47, 53, 66, 188; importance of,
    70–71; monitoring of, 201; photosyn-
    thesis and, 44, 45; phylloxera and, 126;
    structure of, 46–47; temperature and,
    150, 200, 222, 223; types of, 42–43; vas-
    cular system and, 65. *See also* fermenta-
    tion; yeasts

sulfur, 37, 119, 123

sulfur dioxide, 16, 113

Sultana grape, 217

Sumerians, 9

sunlight, 44, 147, 150, 151, 152, 201

"Super Seconds," 144, 152–53

sweetness, 164–65, 167, 181

Swinchatt, Jonathan, 147–49

Syrah, 87

Syria, 8

2,4-D (herbicide), 218

T1r3 taste receptor, 181

Tannat, 127

tannins, 52, 53, 115, 145, 152, 190, 201, 206,
    222, 223

tartaric acid, 7

tartrate, 201

Tasmania, 224–25

taste, 162, 164–65

tautomers, 49

taxation, 16, 21

taxonomy, 75–78

temperance movement, 20, 31

temperature, 12, 139, 146, 150, 198, 199, 200,
    221–22, 223, 224

terebinth, 7, 12

Terral, Jean-Frédéric, 88–89

terroir, 92, 114, 132–54, 175, 203, 204, 205,
    219

testa, of grape, 64

tetrachromatic color vision, 159–60, 161

thale cress weeds, 68

Theophrastus, 102–3

This, Patrice, 82

thylakoid membrane, 44

thymine, 55, 56, 60, 68, 109

Tiedje, Robert, 99

tongue, 164, 165, 166, 167

topography, 151, 152

Torulaspora, 95

Traminer, 87

transcription factors (regulator genes),
    106

transmittance, 160

transpiration, 150
Tremellaceae, 95
Tremellales, 95
Trichoderma harzianum, 218
trichromatic color vision, 159
Tröndle, Dorothee, 79–80
truffles, 93
trunk, of vine, 66
tryptophan, 46
tubers, 29
tumbleweeds, 105
Turin, Luca, 163–64
Turkey, 88
Turkmenistan, 88
Tuscany, 136
twins, 193
Tyson, Neil deGrasse, 34–35

umami (savory) taste, 164, 165
uncompetitive antagonists, 186
Uncorked (Liger-Belair), 168
University of California, Davis, 83, 112, 129, 130, 131, 199
Urartu, 3
urbanization, 8, 20
urine, 183
U.S. Department of Agriculture (USDA), 50–51, 78, 88
Userhat (Egyptian scribe), 13

Vaca Mountains, 147
Vandals, 18
vascular system, 65, 73
vasopressin, 183
Vassal, France, 78, 82, 88
ventral bundle, 64
ventral tegmental area (VTA), 192
véraison, 65, 103
Vespa crabro, 98–99
Vilgalis, Rytas, 94
vinegar, 166
Vins de l'impossible (Huetz de Lemps), 220

Viognier, 87
viruses, 99, 103, 130, 218
Visigoths, 18
vision, 157–60, 188
Vitaceae, 74, 75, 80
Vitales, 74–75
vitamins, 184, 185
Vitis labrusca, 126
Vitis rupestris, 126
Vitis vinifera, 6, 74, 75–76, 78, 79, 80, 81, 124, 127, 129; subspecies of, 76 (table)
Vitis vinifera sylvestris, 77, 80, 81, 88
Vitoideae, 75
volcanoes, 138, 147, 152, 216

Wadem, Nicholas J., 156
Wallace, Benjamin, 207, 210
Washington State, 224
wasps, 95, 98–99
water, 34, 35, 38, 39, 42, 44, 65, 139
wheat, 9
White, Michael, 223
white wines, 53, 54, 87, 161, 190. *See also individual varietals*
wildfire, 223, 225
Willamette Valley, 224
Williams, Roger N., 121
Wilson, James, 145
wind, 104, 133, 138, 150, 151
Wine Advocate, 173
wine critics, 173–77
wineglasses, 165, 167–68
Winemaker's Dance (Swinchatt and Howell), 147–49
Wines (Amerine), 200
wineskins, 14, 19
Wine Spectator, 174
Winkler, Albert, 200
Winogradsky, Sergei, 108
woman suffrage, 21
wood alcohol, 41
worms, 68

xenin, 186
Xenophon, 3
Xinjiang, China, 9
xylem, 65, 130

yeasts, 12, 24–25, 26, 42, 49, 50, 51, 53, 63, 71, 91–100, 115; "captive," 95; on grape surfaces, 52, 111; native, 114; proteins made by, 45; in secondary fermentation, 54, 197; sequencing of, 94–95; structure of, 37, 68, 93, 96, 97, 109; temperature and, 198; wasps and, 98–99, 104; wild, 92, 113–14. *See also* fermentation; sugars
yogurt, 51
Yqem (chateau), 209

Zagros Mountains, 7
Zapater, José Miguel Martínez, 80, 81
Zecca, Giovanni, 79–81
Zinfandel, 172, 224
Zuckerman, Benjamin, 34
zygotes, 69